DRIVING AIRLINE BUSINESS STRATEGIES THROUGH EMERGING TECHNOLOGY

*Dedicated to the memory of my parents
Dharam Pal and Shanti Devi*

Driving Airline Business Strategies through Emerging Technology

NAWAL K. TANEJA

ASHGATE

© Nawal K. Taneja 2002

All rights reserved. No part of this publication may be reproduced, stored in a retrieval system, or transmitted in any form or by any means, electronic, mechanical, photocopying, recording or otherwise without the prior permission of the publisher.

Nawal K. Taneja has asserted his moral right to be identified as the author of this work in accordance with the Copyright, Designs and Patents Act, 1988.

Published by
Ashgate Publishing Limited
Gower House
Croft Road
Aldershot
Hampshire GU11 3HR
England

Ashgate Publishing Company
131 Main Street
Burlington VT 05401-5600 USA

Ashgate website: http//www.ashgate.com

Reprinted 2002

British Library Cataloguing in Publication Data
Taneja, Nawal K.
　Driving airline business stategies through emerging technology
　1. Airlines - Management 2. airlines - Planning 3. Airlines - Technological innovations
　I.Title
　387.7'0684

Library of Congress Control Number: 2002102526

ISBN 0 7546 1971 0

Printed and bound in Great Britain by MPG Books Ltd, Bodmin, Cornwall

Contents

List of Figures	*viii*
List of Tables	*x*
Foreword	*xi*
Preface	*xiii*
Acknowledgements	*xv*

1	**Challenges Driven by the Changing Airline Customer and Industry Dynamics**	1
	The Changing Airline Customer	1
	Demographic Trends and Implications	1
	What do Customers Want?	10
	Industry Dynamics	14
	Competition	14
	Technology	18
2	**Opportunities Driven by Emerging Technology**	27
	Operational Planning	28
	Products Development	37
	Aircraft	37
	On the Ground	42
	Challenges	48
	Airline Profitability Measurements	50

3 Market Segmentation and Customer Relationship Management   57

Market Segmentation   57
  Evolving Strategies   58
  Segmentation Techniques   63

Customer Relationship Management   72
  What CRM is and What it is not   72
  Critical Success Factors   75

4 E-Business and its Application to Airlines   85

Overview of E-Business   87
  Description of E-Business and its Evolutionary Stages of Development   87
  E-Business and its Effects on Four Constituents   92

Key Drivers of E-Business   99
  Vision   100
  Technology   100
  Processes   104
  Metrics   105
  Transformation Process   106

5 Opportunities Driven by Emerging Aircraft Technology   109

Long-Haul International Markets   109
  The Airbus A380   109
  The Boeing Sonic Cruiser Concept   115

Regional Markets   118

Personal Transportation   122
  Fractional Aircraft Ownership   123
  Personal Aircraft   126

6 Forces Transforming the Air Cargo Market	131
Changing Needs of Shippers	131
Evolving Industry Structure	136
Independent Units of Combination Carriers	136
Consolidation and Blurring Boundaries	137
Bridling Technology to Redefine the Air Cargo Market	145
Aircraft	145
Information Management	147
7 Business Structures and Processes to Capitalize on Emerging Technology	153
Vision and Role of Technology	154
Is Technology an Enabler or Driver of Business Strategy?	154
Basic Technologies or Exotic Technologies?	158
Intelligent Assessment of Technology	160
Business Philosophies, Structures, and Systems	163
Value-Based Planning	163
Chief Information Officer	168
Legacy Systems	170
Integrated Databases	172
Knowledge Management	176
8 A Call for Action	185
Index	*193*
About the Author	*197*

List of Figures

1.1	World populations, 1950-2150, United Nations "Medium" projection	2
1.2	21 Largest cities in terms of populations in 1950	3
1.3	21 Largest cities in terms of populations in 2000	3
1.4	Potential for growth in air travel	5
1.5	Real GDP 1999 versus populations	8-9
1.6	International passenger yield trend	15
1.7	Competition among airlines and trains within Great Britain	17
1.8	Changing industry and customer dynamics and the role of technology	25
2.1	virtuGate passenger hours and impacts on cash flow	36
2.2	Economics of first class indexed to economy class	39
2.3	Economics of business class indexed to economy class	39
2.4	Economics of premium economy indexed to economy class	40
2.5	Potential product innovation on ultra long-haul sectors	41
2.6	Simplifying passenger travel project process flow	45
2.7	Variation in the profitability of economy-class passengers	52
2.8	Segmentation profitability	53
3.1	A value-based segmentation relating to loyalty program status	61
3.2	Product development based on need brand relationship: Southwest Airlines	64
3.3	Four major components of a CRM initiative	73
3.4	Alternative methods for implementing a CRM initiative	76
3.5	CRM—an evolutionary process	83
4.1	Airline industry triangle: relationships, challenges, and trends	88
4.2	Value of e-business	89
4.3	Nine factors drive overall airline satisfaction	98
5.1	Airbus A380	110
5.2	Top ten airport-pairs in 2019	111
5.3	Cabin configuration possibilities	113
5.4	Boeing's proposed Sonic Cruiser	117

5.5	Penetration of regional jets in hub-and-spoke and point-to-point markets	119
5.6	Cross sections of jets for regional markets	121
5.7	Eclipse 500	128
5.8	Cirrus SR-20	129
6.1	Passenger and freight yield trends	132
6.2	Air cargo service providers and one interpretation of their business models	141
6.3	Air cargo service providers and a second interpretation of their business models	142
7.1	Cross-functional integration	165
7.2	An integrated data system	174
8.1	While passenger traffic growth has outperformed GDP growth, real revenue growth has not	185
8.2	Value versus investment/integration required for selected technologies	189
8.3	CEO cockpit executive information system	191

List of Tables

2.1	Why crew scheduling is so complicated	33
3.1	Market segmentation criteria for airlines	59
3.2	Examples of business traveler statements	65
3.3	Examples of clusters based on the TRC analyses	66
3.4	Examples of leisure traveler statements	67
3.5	Insights from leisure traveler statements	67

Foreword

As in any sector, leadership and the use of technology are critical business imperatives for CEOs and other executives who aim to succeed in the airline industry. However the dynamic, growing complexity and extreme competitiveness of the operating environment make it difficult to identify accurate, timely and high-quality information. Pressed to strike the right balance among the main stakeholder groups, airline officers find it increasingly difficult to respond effectively to rapidly changing business and customer environments.

Ever since the first scheduled airline flight in 1912, only a handful of airlines have been profitable and successful. A few airlines have enjoyed consistent success while others have had rocky rides. Indeed, at IATA's 2001 AGM in Madrid, I stressed that airline shareholders are not well rewarded and that "it was ironic that businesses which depended on airlines for their very existence all seem capable of generating better profitability."

Driving Airline Business Strategies through Emerging Technology is a book for senior airline executives on how to harness the power of emerging technology to achieve growth that is profitable and sustainable. It shows practical examples of how technology—ranging from mobile communications and the Internet to the aircraft itself—can be used to reduce costs, enhance revenue, improve customer service and improve security.

The book is based on Nawal Taneja's participation in numerous discussions with a wide range of practitioners in the airline industry and related industries, as well as his own experience in the implementation of emerging technologies in a variety of functions within the airline industry. While there are numerous experts in every area of technology discussed in this book, Nawal Taneja's contribution lies in synthesizing the pragmatic aspects of technology that are applicable to our industry and valuable to a broad range of senior executives within airlines—regardless of their size or type of operation.

To excel in the business environment by harnessing the power of technology requires more than just the decision to acquire technology. It also requires a significant change in attitudes to the role of technology, organization structures, planning philosophies, and culture. The time has now come for our industry, according to Nawal Taneja, "to have a better business discipline to capitalize on emerging technology to generate sustainable shareholder value for *itself*." This is just a different way of delivering the message I gave at IATA's 2001 AGM in Madrid. Hopefully, this book and IATA, through its Information Management Committee will continue to support industry technology initiatives on security and infrastructure, to help airline management navigate profitably through these turbulent times.

Pierre Jeanniot, o.c.
Director General and CEO
International Air Transport Association

Preface

Emerging technology—like mobile communications, information management, the Internet, biometrics, or new kinds of aircraft—can bring significant business efficiencies across a broad spectrum of an airline's operations. It can add value to different functions by reducing costs, enhancing revenue, and improving customer service, as well as customer safety and security. This book is intended to provide the reader, a practitioner in the airline or airline-related industry, with (1) an overview of the impact of emerging technology on the airline industry, (2) a conceptual framework for developing and aligning business and technology strategies, and (3) an appreciation of the organizational changes required to gain the most benefit from the implementation of leading technologies.

This book is not about explaining new technology in detail, nor is it about discussing the airline industry; rather, it is a book on the use of emerging technology within the airline industry. In addition, it not a dissertation on any one area; it is an overview of a wide range of technologies and their actual and potential value for the airline industry. As such, the airline marketing, financial, and operating practitioners may find the description of technical trends useful to formulate their business strategies. Likewise, airline information and technical practitioners may find descriptions of the business challenges and opportunities useful for developing and aligning their technology strategies with business strategies—in other words, an understanding of the business context to which emerging technology must be aligned. At the same time, some technology suppliers might benefit from understanding the complexity of core processes of the airline business and its market dynamics. Such an understanding could be useful for developing and marketing cost-effective innovations based on emerging technology—for example, through the willingness to share financial risks.

The book is divided into eight chapters that discuss dynamics in the airline industry and how emerging technology can help shape the future of this industry using best practices examples in a variety of areas. The first chapter highlights two major forces (demography and customer expectations) that are changing the airline customer, and two major forces (competition and technology) that are changing the dynamics of the industry. Chapter 2 presents some areas of technology that can help in the optimization of operations, development of products, and measurements of customer and product profitability. The third chapter discusses technology that can enable airlines to identify the most profitable customers as well as the appropriate strategies to use in developing relationships with these customers. Chapter 4 highlights some areas in which e-business can assist airlines in increasing their operational efficiency, improving customer service, and boosting competitive advantage.

The fifth chapter discusses the role of aircraft technology helping the airline industry adapt to such challenges as globalization, liberalization, constrained infrastructure, and changing and rising customer expectations. Chapter 6 highlights some key forces—changing needs of shippers, the dramatically changing industry structure, and aircraft and information technology—transforming the cargo component of the airline industry. The seventh chapter presents a number of changes required in airline business structures, processes, and philosophies to capitalize on emerging technology. The final chapter presents one interpretation of how technology has contributed to the change in the landscape of the airline industry and how technology can be used as a resource to proactively manage the inevitable structural changes.

Acknowledgements

I would like to express my appreciation for all those who make this book possible, including those from: Airbus—Adam Brown, Paul Clark, Guy Dallery, Thomas Fouche, David Jones, and Didier Lenormand; Air Canada—Jim Hunt (recently retired); American Airlines—Scott Nason; America West Airlines—Jim Oppermann;—Atraxis—Armin Meier and Linda Stichelbaut; Boeing—Tom Crabtree; Bombardier Aerospace—Chul Lee; British Airways—Rod Muddle (recently retired), Andrew Sentance and Richard Wyatt; Continental Airlines—William Brunger; ExecutiveJet—Richard Smith III; Fairchild Dornier—Carl Albert and Mike Miller; Hewlett-Packard—Alan Nance; IATA—Karin Gebert, Harry Govind, and Mike O'Brien; IBM—Declan Boland; Kale—Ashish Malhotra; Lufthansa—Gabriela Maria Kroll and Oliver Sellnick; McKinsey—Lucio Pompeo; J.P. Morgan—Andrew Lobbenberg; NCR—Steve Dworkin, Brendan Hickman, David Kramer, David Schrader, and Jan Wood; Roland Berger—Michael Beckmann; South African Airways—Andy Hayward; Southwest Airlines—John Jamotta and Pete McGlade; Swissair—Dani Weder, Michael Landthaler, and Volker Heitmann; Taylor Nelson Sofres (Travel Research Center)—Carolyn Childs; US Airways—Ben Baldanza; Virgin Atlantic—Barry Humphreys; and VISA International—Sonia Reed.

There are a number of other people who provided significant help in such areas as the design and formatting of all the exhibits (Benjamin Kann and Ryan Leidal at the Ohio State University), design of the jacket for the book (Araceli Guenther at INTAG), research regarding some material presented in Chapters 1 and 5 (Aya Nakamachi at the Ohio State University), editing the various chapters (Robert Powers, Carolyn Taneja, and Juliet Williams) and production of the book (John Hindley, Consulting Editor, and the Ashgate in-house publication staff, especially Pauline Beavers, Andy Jones, Pam Park and Adrian Shanks.

Chapter 1

Challenges Driven by the Changing Airline Customer and Industry Dynamics

The airline industry is being radically transformed by the intersection of many forces—globalization, liberalization, privatization, hyper-competition, shortening technology cycles, changing demographics, economic and social cultures, rising consumer expectations and power, environmentalism, and the growing concern for safety/security. The first section of this chapter highlights just two of these major forces—demography and customer expectations—that are changing the airline customer. The second section deals with two other major forces—competition and technology. Both are changing the dynamics of the airline industry. A brief assessment of these four forces—two that affect the customer and two that affect the industry dynamics—can help airline managements identify, develop, and implement short- and long-term business strategies, particularly with respect to the role of emerging technology. That is the theme of this book.

The Changing Airline Customer

Demographic Trends and Implications

Demography will play a major role in the development of the commercial air transportation industry in the coming decades. Populations and their key attributes—shifts in populations, distributions of income and ages, and migrations patterns—are expected to have a profound effect. Consider the changes in the distribution of populations between the developed and the less developed countries. See Figure 1.1. Recently, the ratio of populations of developed countries to developing countries was approximately one to five. In the next 50 years the same ratio may reach one to ten. Moreover, the trend is not just that the populations of developing countries are increasing at much higher rate than the populations of developed countries. Populations of some developed countries are in fact decreasing, while some others staying flat, at best. The

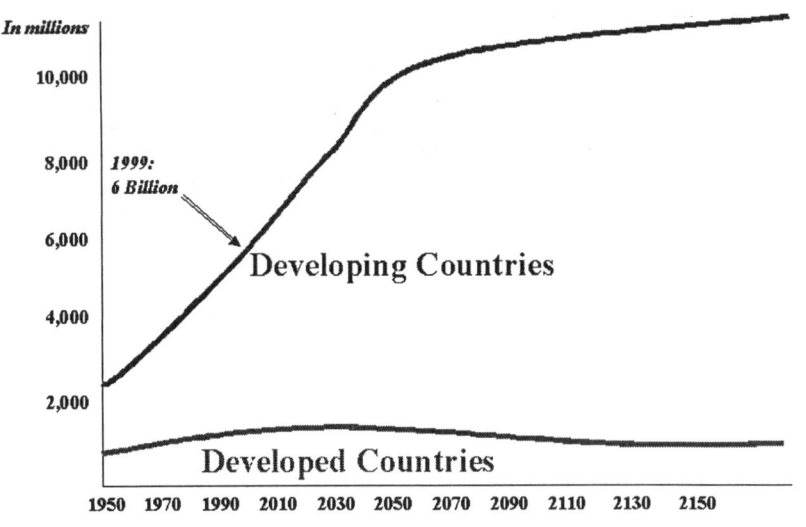

Figure 1.1 World populations, 1950-2150, United Nations "Medium" projection

Source: United Nations and the U.S. FAA 24th Annual Commercial Aviation Forecast Conference Proceedings, March 1999

projection in the decline of populations in developed countries is based on lowered birth rates and is net of the increase in populations due to immigration, for example, the emigration to Western Europe from Eastern Europe.

Figure 1.2 shows the top 21 cities in the world in terms of population in 1950. It should be noted that 12 of these cities were located in developed countries and nine were located in less developed countries. The size of the circle represents approximately the size of the population in relative terms. Notice the tiny size of such cities as Mexico City and Sao Paulo compared to London and New York City. Now look at Figure 1.3 that shows the populations of the largest 21 cities in the year 2000.

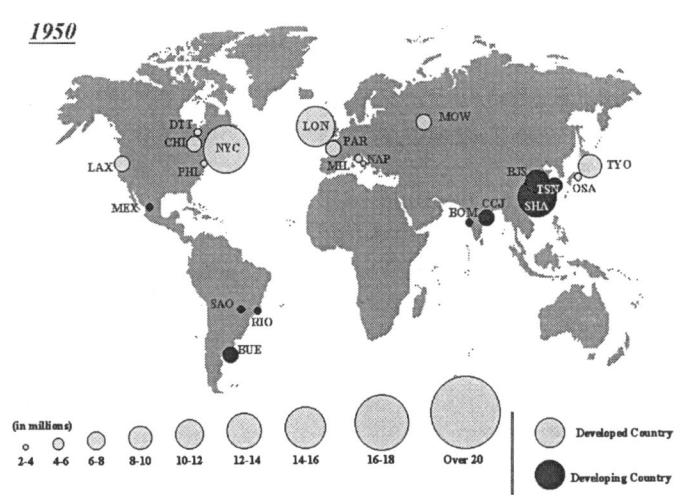

Challenges Driven by the Changing Airline Customer and Industry Dynamics 3

Figure 1.2 21 Largest cities in terms of populations in 1950
Source: Cities of the World: World Regional Urban Development

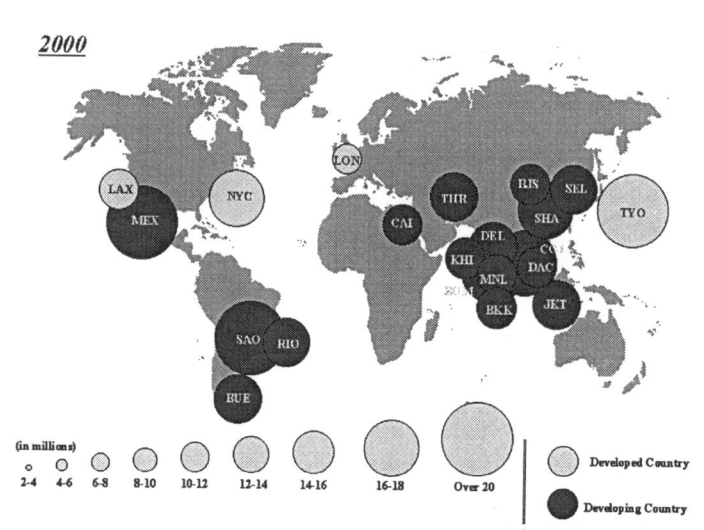

Figure 1.3 21 Largest cities in terms of populations in 2000
Source: Cities of the World: World Regional Urban Development

This time, 17 of the top 21 cities (again in terms of populations) are located in developing countries and only four in the developed countries. In fact, if the top 20 cities had been selected instead of the top 21, the list would not have included even a single city in Europe. London was the 21st largest city in terms of population. Notice also now the size of Mexico City and Sao Paulo. Two other attributes about Tokyo are note worthy. First, it is the largest city in the developed part of the world (according to the 1950 definition of the developed region). Second, it is the only city that is large enough to be in the ranks of the largest cities in this list such as Mexico City and Sao Paulo.

The airline industry may see significant growth to and from, or within selected regions of the less developed parts of the world. The types of products demanded by passengers in less developed regions of the world are likely to be very different from the types of products demanded in the developed regions of the world.

Population alone is not a sufficient indicator of the ability to travel by air. Populations must also have the economic means to travel by air. However, even if a very small percentage of a very large population has the economic means to travel, it could amount to a significant proportion of the air travel market. Consider the case of Sao Paulo. Obviously, there must be a significant number of passengers traveling between Sao Paulo and Miami to justify six daily nonstop flights with wide-body aircraft in each direction during a typical week day in January 2002.

Figure 1.4 shows, from a different perspective, the huge potential for air travel growth relating to less developed countries. There are a number of points worth noting about the information contained in this figure. It is a plot of trips (by air) per capita and real gross domestic product per capita in 1997 US dollars. First, for a number of countries above the curve such as Bolivia, Malaysia, New Zealand, Ireland and Iceland, air travel is more convenient than ground transportation due to a broad spectrum of barriers such as mountains, bodies of water, lack of adequate roads and railroads, and so forth. In such countries, air travel is much more convenient. Second, at one end of the scale, while in the United States each person may be making three or four trips a year, at the other end of the scale, in some developing countries (such as India), it could be one person out of every 100 making one trip per year. So an increase in that later number to just one person in 10 making one trip per year could result in an enormous increase in the total travel by air.

In the United States the average person making three or four trips per year may not only be unwilling to make any more trips but could in fact make one less trip per year if the hassle associated with travel increases or if alternatives become more viable, such as an increase in the quality and availability of video conferencing (and Web conferencing) and a decrease in its cost. Such alternatives are not expected to have a large impact on air travel but they could have a small impact in certain situations such as intra-corporate travel that may not require a face-to-face interaction. For the people in the less developed countries, air travel, on the other hand, is not a hassle but rather is a dream. And for that small segment of the population that has already met its basic needs of food and shelter, people may be willing to spend a larger percentage of their income on air travel compared to the people in the developed countries.

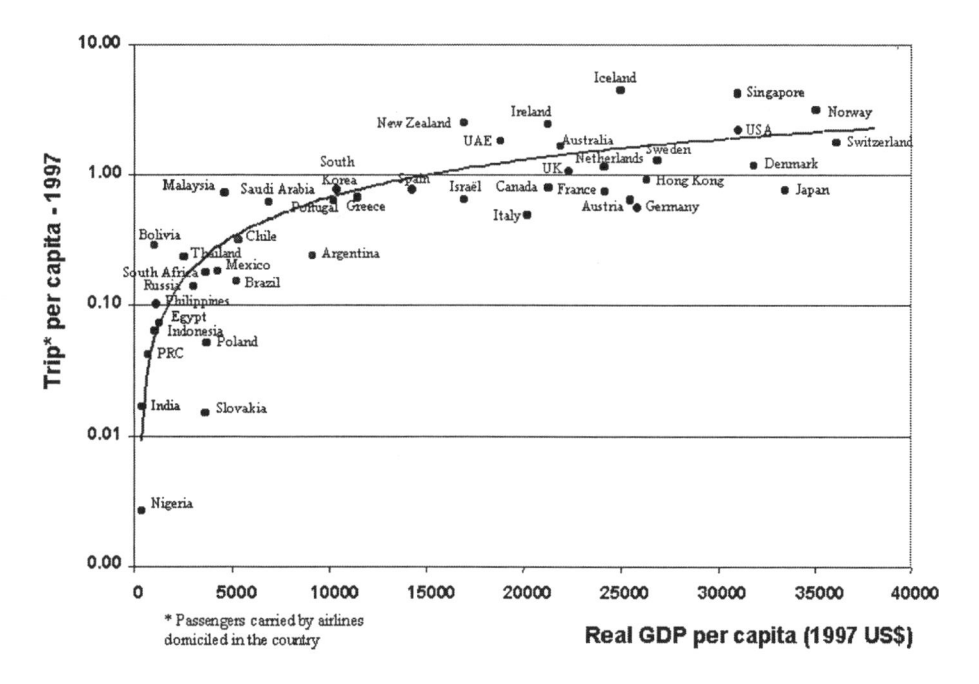

Figure 1.4 Potential for growth in air travel
Source: AIRBUS, ICAO, and Standard & Poors

According to the shape of the curve displayed in Figure 1.4 a small percentage increase in the economies of less developed countries such as India and China is likely to produce a disproportionately large increase in the amount of air travel for reasons mentioned above. In the developing parts of the world air travel may be a higher priority once other basic needs have been met. Moreover, other alternatives for travel may not be viable.

All developing countries shown in Figure 1.4 have not been able to take advantage, to the same degree, of the globalization process—spread of international trade, financial markets, foreign direct investments, emerging technologies, advanced communications, and global patterns of production and consumption. Examples of economies that have benefited the most from the globalization process are located in the Asia-Pacific region (India, China, and Indonesia) and Latin America (Brazil and Mexico). This economic growth has resulted in an increase in travel. Just consider travel between Beijing and some other large cities in China—Hong Kong, Shanghai, Guangzhou, Shenzhen, and Xian. They are already generators of large amounts of air travel and could become much more significant in the next 20 years.

Different strategies have penetrated the air travel market in less developed countries vs. developed countries. In the United States, the air travel market is becoming saturated. At one end of the spectrum, airlines are reducing fares further and further to encourage travel at the low end of the market. At the other end, airlines are either seeking more and more premium passengers though higher and higher quality of service or attempting to "sell up" the service to obtain better yield in return for higher value. However, in developing countries where air travel may be a dream for some segments of the marketplace, the product demanded will be very different—for example, a basic seat in a high-density configured cabin, or low-frequency, or less convenient schedules. The second aspect of the product relates to frequency. The low end of the marketplace will travel whenever the airline schedules the aircraft. The desire for high frequency is not the same in all markets and once a certain number of flights have been scheduled the value of additional flights provides diminishing returns. Finally, even in developing countries there will be a need for a small amount of premium travel. In some cultures, there will always be small segments that will spend the money to go first class, either for comfort or for prestigious reasons.

The four parts of Figure 1.5 portray a different viewpoint of the changing nature of populations in four different regions of the world. Each illustration is a logarithmic plot of the real gross domestic product and population. The first two provide the information on Western Europe and Central Europe/CIS. The next two show the information for the Asia-Pacific and Latin America regions. Each chart has three lines representing three levels of incomes, 200, 2000, and 20,000 dollars per capita. Once again, the real potential for significant growth in air travel is not within North America (not shown included in the chart) or Western Europe but within Central Europe, Latin America, and the Asia-Pacific regions. And within these groups, the real potential is for countries projected to have high growth in income per capita—such as those located in East Asia. Unfortunately, most countries in Africa (not shown here), particularly Africa Sub-Sahara, will continue to be a tiny part of the air travel market.

If we consider the growth and shift in populations as one important component of demography, the impact of immigration is the second important component of changing populations. This component relates to an increase in the number of immigrants to the developed countries, partly for greater economic opportunities and partly for political freedom. The United States has been experiencing a significant growth in the Hispanic segment of the population. During the first decade of the 21st century the Hispanic segment of the U.S. population may represent the largest minority, surpassing the African American segment. Moreover, although small in base, Asians now represent the fastest growing segment of the U.S. population. Within this segment, higher levels of education could also imply significant gains in personal wealth and purchasing power for this minority group. The impact of this segment (with respect to its purchasing power) will not only be an increase in the amount and type of goods and services purchased, but also the brand loyalty relating to the goods and services purchased. Similar trends are being experienced in other developed countries such as those in Western Europe and in Japan where, for example, ethnic Japanese from Brazil and Peru are moving in to fill manufacturing jobs as well as to take care of the growing numbers within the elderly segment.[1]

Besides the growth/shift in populations and impact of immigration, other changes in demography include, the emergence of a large middle class in selected developing countries, a significant increase in the number of women entering the workplace (especially in developed countries), and

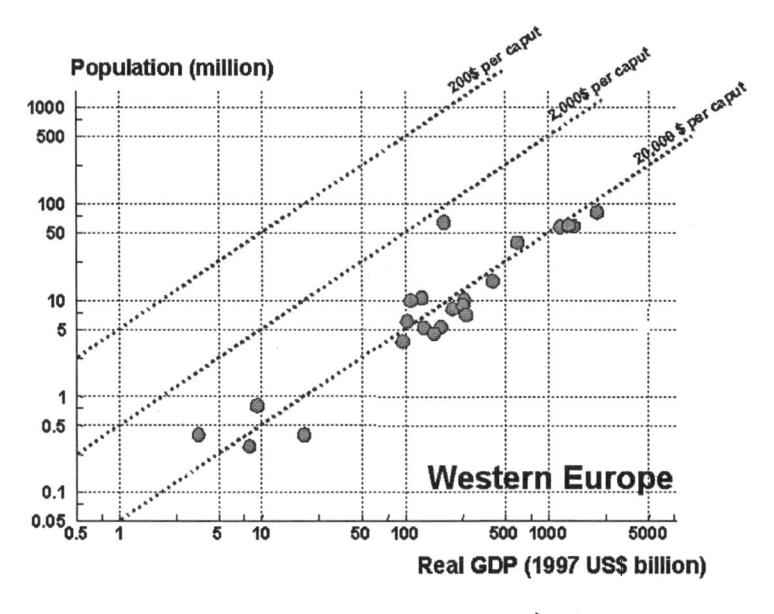

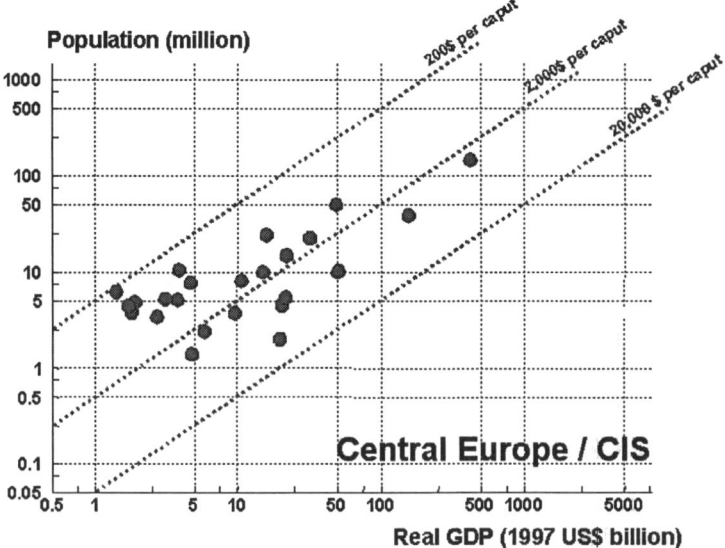

8 *Driving Airline Business Strategies through Emerging Technology*

Figure 1.5 Real GDP 1999 versus populations
Source: AIRBUS

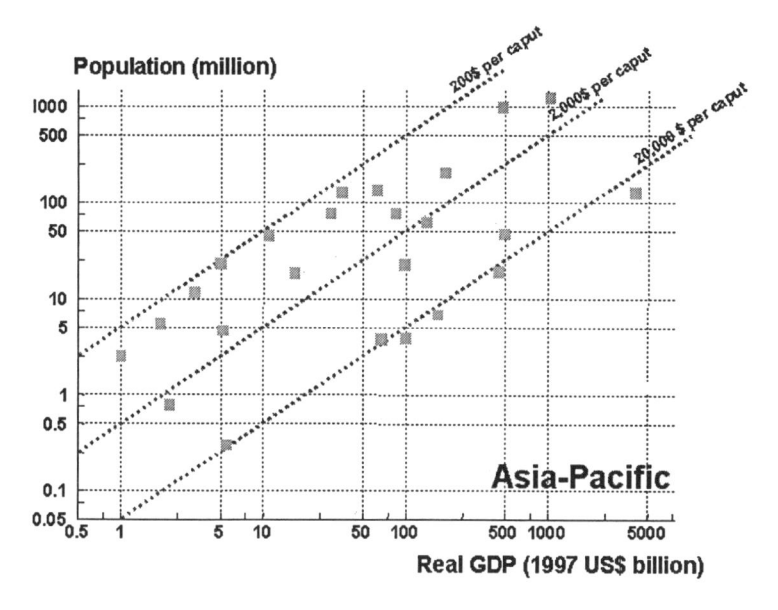

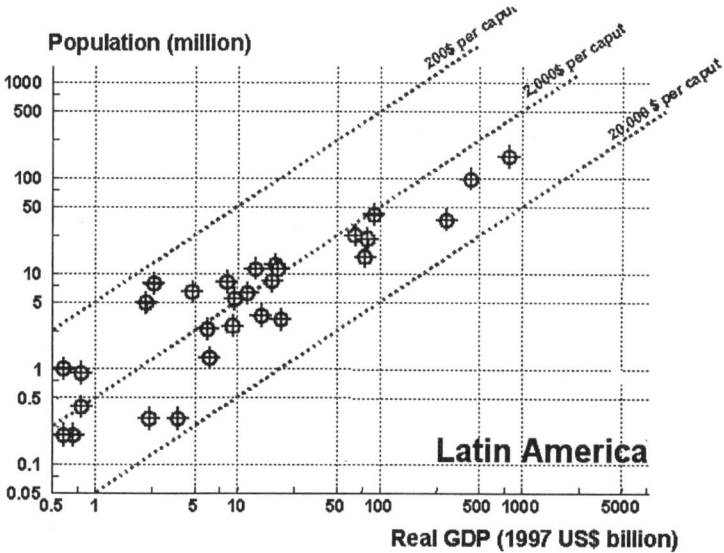

Challenges Driven by the Changing Airline Customer and Industry Dynamics 9

Figure 1.5 (continued) Real GDP 1999 versus populations
Source: AIRBUS

an increase in life spans resulting from such factors as advances in medicine, health care and education. All these trends will have varying degrees of positive influence on air travel.

What do Customers Want?

A number of forces are changing the expectations and power of customers. The most important forces appear to be huge amounts of information and the availability of communication technologies to make this information available to virtually any one, at any time, and at any location. And it is the availability of this information that provides customers empowerment.

Customers now have access to extensive information on competitors, products, availability, and prices made possible through the Internet. Moreover, the Internet has made it relatively easy to switch competitors. A passenger can now find out easily the lowest fare for a New York-London trip in business class or who offers certain types of services such flat beds, meals on the ground prior to departure, and ground facilities at the arriving airport for freshening up.

Customers, at least in some regions of the world, expect business process transparency with respect to prices and services, for example, on time performance of flights, penalties associated with restricted fares, the amount of care the airline will provide to a passenger on a given fare in case of problems such as missed connections requiring overnight stay, meals, ground transportation, and long-distance telephone calls. They expect to be informed about inventory. How many seats are available at the low fares advertised in newspapers? Will they be able to redeem their mileage awards without unreasonable conditions? What are the policies regarding promises made by travel agents? Basically, customers want to make better decisions to make their lives more pleasant or, at the very least, predictable.

Many passengers do not understand the rationale of the airline industry's fare structure and fare levels, particularly since the introduction of revenue management systems. Although they understand the difference in fares among the different classes of services (first class, business class, and economy class), they do not understand the level of these differences. The confusion appears to be first in the differential between an economy-class fare and a business-class fare, for example, for a trans-Atlantic flight and second in the differential between the deeply discounted but heavily restricted economy-class fares and the normal economy-class fares.

Consider the information made available to a passenger on the website of a brand-name airline for a normal economy-class fare and a business-class fare for a round-trip between New York and London. A passenger wanted to make the reservation on 28 December 2001 for outbound travel on 02 January 2002 with a return on 05 January 2002. The fare quoted on the website was US$1,199 for economy class and the US$7,380 for business class. The general public is fairly well aware of the perks and extra services provided to business travelers such as limousine service to and from the airport, special check-in counters, upgrade to premium-class cabin, additional space in the cabin, superior class and a choice of meals, a comprehensive in-flight entertainment system, and airport lounge facilities at both ends of the trip. However, passengers have difficulty in placing a value for each of these service attributes to explain, in this case, a difference of US$6,181.

Another passenger for the same itinerary and travel dates received similar type of information for travel in the economy class on a ticket that is restricted and one that is not restricted. The non-restricted fare was US$1,199 and the restricted fare was US$350. The first restriction was that the earliest day the passenger could travel to London would be 07 January 2002 and the earliest date of return would be 14 January, 2002.

Cancellation charges and fees for changing the reservation are two other examples of restrictions on discounted fares. In this particular case, it was even possible to obtain a fare as low as US$290 if the passenger was willing to travel on 09 January outbound and return on 15 January as well as travel in and out of Newark instead of JFK Airport. Again, passengers understand that they need to pay for flexibility and elimination of other restrictions, but they are not able to associate the differential amounts for each type of restriction. Moreover, even lower fares were available through other airlines and other channels of distribution. Consequently, although from an airline's point of view it makes good sense to implement revenue management systems to compensate for the complexities of the air travel business (such as variability of demand, perishability of the product, and existence of a broad spectrum of price-elasticity of demand), passengers have difficulty understanding not only the differential in economy-class fares for, according to their perception, virtually the 'same product' but also the reason for constant change in fares. One explanation is available in the next chapter relating to revenue management. From the general public's viewpoint airlines need to make fare structures and levels less mysterious, less complicated, and with a full explanation of options and costs of options.

Customers expect relevant, accurate, consistent, and timely information so that they can have some control during irregular operations. Airlines are already beginning to provide all kinds of information to their top end of business travelers—information that has higher value than free tickets and upgrades. Some airlines have begun to send to their best customers, via a broad spectrum of mobile devices, information on flight delays, causes of flight delays, flight cancellations, and available alternatives. Moreover, passengers want the information to be linked. If a flight is late, they want the information relayed to car rental agencies, hotels, and so forth. This expectation is particularly important for the movement of cargo that depends on multiple parties. Delays or problems at one location must be transmitted immediately to all parities. See the discussion in Chapter 6.

This information-related value-added service begins to address one of the major expectations of top-end business travelers. An interesting aspect of this service is that it could be provided free to higher-margin customers and on a fee basis to lower-margin customers. An airline could collect specific customer data on the type of information needed, the mode of delivery desired, and the willingness to pay a fee for services provided. Some passengers may be willing to pay for vital information such as serious delays and cancellations while others may be willing to pay for simple conveniences—for example, the closest parking lot to an airline's check-in area that still has empty parking spaces and the location of such spaces.

Emerging technology—particularly in the area of the Internet, wireless communications, and information management—is raising the expectations of customers. Some customers now expect communications to be personalized, not just with respect to mode—such as e-mail vs. telephone vs. pager—but also with respect to their historical relationship with the service provider. Moreover, some customers are even becoming aware of the emergence of intelligent services that are raising their expectations even more. Intelligent services relate to the desire of passengers to have access to services that add value in terms of making their environment more productive and their life less stressful and more meaningful. The service could be provided by a person or a software program involving such aspects as finding the appropriate travel itinerary to suit the expected needs of the customer and resolving any problems that may occur to the expected satisfaction of the customer. The feasibility of intelligent services depends on four critical success factors, the Internet, mobile communications, interconnected devices, and interconnected services and service providers. The Internet and mobile communications are already here although the application and standards of mobile communications vary worldwide. The missing links—the lack of connected devices and services—are not far behind. Blue tooth, for example, replaces hardwire links among devices with wireless links providing inter-operability among different devices and systems—even those produced by different manufacturers.

Throughout this book mention is made of different ways of achieving a sustainable competitive advantage. Understanding customer expectations is just one of these methods. An understanding of customer expectations will affect not only the strategy relating to the standard elements of the marketing mix but also other important areas such as customer loyalty and customer segmentation (discussed extensively in Chapter 3).

Customer loyalty has become more important and more threatened as a result of heightened competition. Moreover, the increase in competition has increased the cost of loyalty programs that provide a wide variety of benefits to loyal customers, such as free tickets, upgrades to premium cabins, separate reservation telephone numbers and check-in counters at airports, use of airport lounges, special boarding priority and seating preferences. In order to meet customer expectations, airlines need to track not only the value provided and received from these loyalty benefits but also the costs of providing such benefits. Conventional wisdom implies that the cost of retaining existing customers is generally lower than the cost of acquiring new customers. However, conventional wisdom may not be valid for all airlines and in all situations. While the cost of serving a loyal customer may be less in some aspects due to the fact that the customer gets to know the airline and its product line, the cost of keeping some customers loyal may be very high due to constant competitive threats. Moreover, as discussed in Chapter 3, not all loyal customers are profitable.

As the drivers of customer loyalty vary from segment to segment, customer to customer, phase of the trip to phase of the trip, and market to market, it becomes important to understand customer expectations at a granular level. Obviously, all customers have the same expectation with respect to the basic attributes of the product such as safety, reliability, and a competitive price. However, additional expectations vary by some of the factors listed above. And it is the responsibility of the airline to identify customer expectations and then select the segments of the marketplace whose expectations it can meet profitably and consistently with its value propositions. Moreover, it is necessary to control and shape the expectations of customers based on their value. For example, low-margin customers calling reservation centers can be routed to automated voice response systems while high-margin customers can be routed to live agents and even higher-margin customers can be routed to dedicated agents to provide truly customized, personal, and interactive service. The key elements of the analysis of customer expectations are selecting customer segments whose expectations fit strategically with the value propositions of the airline, meeting the expectations of all customers in the selected segments, and controlling and shaping the expectations in line with the branded value propositions. Southwest Airlines is probably the best example of an airline that comes closest to meeting these criteria. For full-service airlines with worldwide networks and strategic alliance partnerships, technology is available to assist them in matching (strategically, profitably, and consistently) the changing customer expectations of selected segments and an airline's value propositions.

Industry Dynamics

Competition

The impact of globalization in the airline industry has led to a significant progress toward the liberalization of air services and an increase in competition within the airline industry. In the United States, deregulation of domestic markets resulted in a significant decline in passenger fares during the past 20 years. During the year 1980, when the full-fare yield was 15 cents, 57.5 percent of the passengers traveled on discounted fares, and the average discount was 42.9 percent of the full fare. During the year

Challenges Driven by the Changing Airline Customer and Industry Dynamics 15

2000, while the full-fare yield had increased to 43.81 cents, 94.6 percent of the passengers traveled on discounted fares, and the average discount was 71.1 percent of the full fare.[2] The average fare paid by passengers in U.S. domestic markets has been declining in real terms. A similar trend can be observed in international markets. See Figure 1.6.

In Europe, the liberalization process has led to a common civil aviation market. This means that once an airline has received appropriate safety and financial clearances from the government authorities in a European country in the Union, the airline can operate services between any two airports in the European Union, exemplified by the operations of easyJet, Ryanair, and Virgin Express. In the case of Australia and New Zealand, governments have even permitted airlines with foreign ownership to provide service in domestic markets as well as the control of their domestic airlines. Virgin Blue is an example of an Australian airline that is foreign owned and that offers low-fare services within domestic markets in Australia.

Real international yields*

*US cents per RPK, scheduled airlines, deflated by US CPI, 1990=100

Figure 1.6 International passenger yield trend

Source: British Airways

After the United States, the impact of increased competition is most visible within Europe in general and within Great Britain in particular. In recent years the low-fare airlines have been growing at a much faster rate than the large traditional airlines. In Great Britain, for example, low-fare airlines have grown by approximately 40 percent in the past five years. This growth has taken place almost entirely at the smaller regional airports. Between 1995 and 2000, passenger traffic grew by about 19 percent at London's Heathrow airport compared to more than 200 percent at Stansted, Luton, and Liverpool.[3] Part of the explanation for the high growth of low-fare airlines in Great Britain is that their fares are not only considerably lower than traditional airlines but in some cases they are also lower than train fares.

For reservations made on 27 December 2001 for travel the next day on 28 December, 2001 from London to Glasgow, easyJet and GO offered fares that were about 40 percent of the fare available on British Airways and about 50 percent of the fare available on British Midland. It is interesting to note that in this example the fares offered by easyJet and GO were even lower than the standard fares available on trains (Virgin Trains and GNER). See Figure 1.9 that is based on the example contained in the June-July 2001 issue of the Economist. This information shown in Figure 1.7 is updated and extracted from the websites of individual companies. It is interesting to note that in this market there are three airlines that offer fares comparable to or less than trains that can take between five and six hours compared to airlines that take between 70 and 90 minutes for the trip between London and Glasgow. Train times are city center to city center whereas airline times are airport to airport that need to be adjusted for lengthy check-in times at airports.

It is also interesting to note the composition of passengers traveling on low-fare airlines. According to easyJet, a significant portion of its passengers are traveling on business, based on their travel behavior (for example, time of making the reservation relative to the day of travel and the day of week of travel). This information signifies that business travel is price sensitive, or at least some portion of it. According to one analysis, the low-fare airlines in Britain have not only lowered the price of air travel but they have also changed the life style of at least some British people—how they travel, where they live, where they work, and where they vacation.[4]

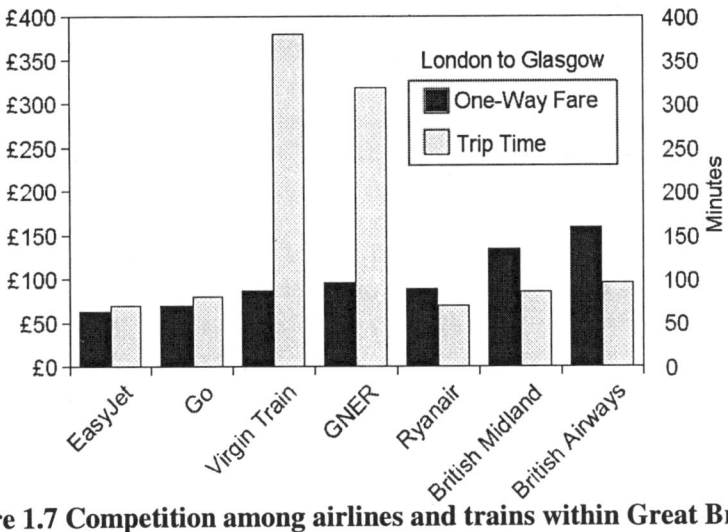

Challenges Driven by the Changing Airline Customer and Industry Dynamics 17

Figure 1.7 Competition among airlines and trains within Great Britain
Source: Websites of Individual Companies

While deregulation is transforming the airlines from slow-moving and reactive virtual monopolies to fast-moving and proactive competitors, the Internet is also accelerating the intensity of competition within the airline industry by enabling airlines to maintain an online global presence. Consider the power of the Internet in enabling easyJet to sell almost its entire inventory online. Consequently, new competitors are entering the marketplace with new business models whether these competitors are actual airlines or new channels of distribution such as Travelocity and Hotwire. In some cases, this has forced the established airlines to assume a defensive position by introducing, for example, low-cost subsidiaries or competitive channels of distribution such as Orbitz. Almost every airline is now trying to reduce its distribution costs. The incumbent airlines cannot, however, rely totally on the Internet. They are attempting to optimize their multi-channel strategy consisting of not only the aggressive use of the Internet but also traditional sources such as travel agents. Competition, direct or indirect, is therefore changing the airline industry's economic dynamics.

Technology

According to some writers in the United States we have moved from industrial economy to information economy (also referred by some as knowledge economy) and are now moving to an experience economy.[5] In the experience economy, consumers' interest is not just focused on buying products or services but also on the total experience consumers receive from the purchase of products or services.[6] Examples include the eating experience that a restaurant may provide or piece of mind that an airline will take care of the passenger if irregular situations occur. In these authors' terms, the product needs to be "experientialized"—the BMW driving experience, the eating experience in a fine restaurant, the reading experience in such U.S. book stores as Barnes and Noble. Such expectations on the part of customers provide businesses opportunities to break out of the commodity problem by differentiating their products or services and at the same time have the potential to charge premium prices for their products or services.

Some aspects of the changes in the structure of the economy have been facilitated by the computer and communication revolutions that have shifted the power from businesses to consumers and from managers to employees. For example, customers can now seek their own information from the business or outside the business and create their own products using the Internet that provides an opportunity to make transactions interactive and personalized. Likewise, Intranets are enabling employees to participate in the decision-making process.

Recent structural changes in the economy started with the introduction of the personal computer in the 1980s which enabled decentralization of computer processing power and decision-making process.[7] At the beginning, personal computers were more or less just sophisticated typewriters and calculators.[8] With time, personal computers became more powerful and provided more valuable applications. Moreover, the favorable price/performance trend relating to the computer chip began to have a tremendous impact on certain products and in some cases the services associated with products.[9] Consider, for example, the increasing amounts of computing power incorporated in cars and the added-value provided by the availability of such computing power.

Then, in the 1990s, individual computers began to be connected through networks. The networked personal computers,[10] coupled with the

Internet, began to communicate with each other leading to a revolution in communications.[11] Next, came such events as the open technological standards, open architecture, and deregulation of the telecommunication industry.

Emerging advances in technology—in such areas as broadband and mobile communications—are enabling consumers to receive information that is both rich and widely distributed.[12] Richness of information refers to its content (quality, quantity, accuracy, timeliness, and customization) and reach refers to the number of receivers of the information. Until recently a business could not communicate information rich in content to a broad group of receivers. A television advertisement can provide breadth but not information that is rich in content. A personal sales representative can provide information that is rich in content but only to a limited audience. Consequently, there was a trade off between richness and reach of the information. Emerging technology removes this obstacle. For example, some passengers believe that they no longer need a travel agent who can provide customized and detailed information about a trip. They can do it on their own using the Internet. Another example would be the ability of an airline to send information-rich (and customized) offers to selected members of its frequent flyers.

Advances in technology have also led to changes in the role of technology in businesses. During the 1980s, information technology played a role that simply supported businesses. In the 1990s, it began to enable key business strategies. In contemporary times, it is actually becoming a part of the business strategies—exemplified by the growth in the e-business. See Chapter 4. Examples of areas where information technology has become part of the business strategy include direct selling of airline tickets through the Internet, the market entry by online service providers such as Expedia, Travelocity, and new marketing initiatives such as new pricing models introduced by such companies as Priceline and Hotwire. Consequently, information technology has moved from supporting back-office functions to the front end, enabling airlines to move from providing standardized customer service to personalized customer service that is cost effective.

Other areas of emerging technology—such as biometrics and voice recognition—have the potential of making travel easier (as described in the next chapter) and cheaper as well as making travel safer. Such technologies can be used to make the validation process more secure and less labor intensive. Prior to the 11 September events, the focus was on convenience and personalized service. Since then the focus has changed to safety and security.

Safety and Security   Using biometric technology to measure human traits—fingerprints, iris and retina patterns, voice patterns—can in fact make the service more convenient (by reducing the length of lines at airports, customs, and immigrations) and at the same time provide improved safety and security for passengers. Biometric cards can contain physical information about the holder (for example, eye shape, eye color, the exact pattern of veins on the retina, and widths and lengths of fingers). These cards can be linked to a palm reader that can measure and compare a broad spectrum of features on the palm of an individual's hand.

Face recognition is another promising biometric technique for busy airports during crowded situations. The technique relies on cameras to identify people at a distance. Basically, a camera creates a digital map of a person's face and then converts the information into a digital code that can then be used to compare the information with the data stored in the system. Consequently, face recognition technology can provide an excellent capability for surveillance because in addition to verifying a person's claimed identity, it can also pick out a person from a database. Moreover, this technology can work in real time by monitoring crowds and movements to identify suspicious persons and suspicious behavior (for example, through monitoring facial expressions). Business intelligence technologies and databases can then be used to analyze behavior patterns.

Finally, technology is available to (a) create new databases and expand existing databases, and (b) integrate the systems to scan hand geometry, face recognition and iris patterns. This technology will verify if the passenger is who he or she claims to be and if the person who boarded the aircraft is the same person who was screened and accepted. Moreover, technology now has the potential to evaluate if a person is a potential risk. The positive identification process can be applied to passengers, employees, third-party handlers and visitors as well as to baggage, cargo, and mail.

Frequent flyers can be pre-screened by providing the necessary information beforehand (for example, finger prints and voice prints) and having the information updated periodically in exchange for convenience. They can then be issued special travel cards containing personal information embedded in a computer chip in the card. However, even though such technology may make it more convenient for frequent travelers to be processed at airports, some may choose not to use this technology because of its impact on issues relating to privacy and civil liberties.

There are two vital requirements for the adoption of emerging technology to improve safety and security. First, in the short-term, many of the recently-introduced systems would need to be modified (self-service machines, wireless devices for check-in, and off-site check-in facilities). Second, there would need to be a linkage among the information systems of airlines, airports, law enforcement organizations, and national and international security agencies. And technology is available to provide such linkages.

Cultural Aspects Although emerging technology has the potential to reduce costs, enhance revenue, and improve service, the degree to which it can be applied will depend on (a) the financial capability and mindset of airline management, (b) differences in customer cultures and attitudes toward technology, and (c) the digital divide. The digital divide is a gap between those with access to information and communication technologies and those without. It is not necessarily a gap between developed and developing countries. In each of these areas, airlines need to assess the costs and benefits of current and emerging technologies from the point of view of their own operations and their acceptance by employees and different segments of the customers they serve. This section provides some information on the last aspect, namely, the potential reaction of different segments of the customer base.

Cultural preferences for the use of emerging technology can be observed relatively easily. For example, e-mail and voice-mail appear to be the dominant mode of electronic interaction within the United States. In Europe and Japan, mobile communications and Short Message Service (SMS)—messages that are normally short and to the point—are more dominant in communications.[13] The penetration of cellular phones is much higher in Japan compared to the United States. Some ethnic consumers are less comfortable with using their credit cards at websites. Smart cards were a success in Europe but not in the United States. Mobile-commerce—enabled by wireless Internet technologies—appears to be more pervasive in Europe and Japan than in the United States. It is important not just to be aware of such cultural differences with respect to the adoption of emerging technology, but also the causes of these differences so that appropriate strategies can be developed to meet the needs and expectations of different segments of the marketplace.

Let us take the first example of the U.S. consumers' preference for e-mail and voice mail and the European and Japanese consumers' preference for SMS. One explanation could be that the U.S. consumer is not satisfied "with small screens, difficult text input, and slow network connectivity."[14] There are two explanations regarding the higher penetration of cellular phones in Japan. First, the United States has been behind the wireless application protocol (WAP) technology adoption curve than Japan due to difficulties involved with its use and the poor availability of the content. Second, some Japanese are observed to use their cellular phones for more non-essential communications—a device for idle chatting as opposed to a device for increasing productivity.[15] The reluctance of some ethnic groups to use credit cards on the Web is based mostly on reluctance to give out personal information.

Here are some explanations for the greater success of smart cards in Europe. Initially, Europe experienced more problems with online verification of credit-card purchases due to poorer telecommunication systems compared to the United States. Therefore, the use of smart cards reduced authorization costs and fraud losses. Second, regular magnetic cards appeared to fulfill all of the customers' and merchants' needs in the United States. Third, the cost of smart cards was much higher than the card with a magnetic strip (about US$3 vs. 25 cents).[16] However, despite the initial success of smart cards in Europe, this technology will become pervasive because smart cards have many more uses than improved authentication. They can hold a broad spectrum of personal information and personal preferences. And they can improve the purchase process, track consumer behavior, and enhance loyalty initiatives.

As to the cultural differences regarding mobile-commerce, part of the explanation for different rates of adoption of this technology is based on the existence of different standards, technologies, and marketing applications. The importance of marketing applications should not be overlooked. Even in Japan, where the demand for NTT DoMoCo's iMode wireless Internet service is increasing at a phenomenal rate, the potential is even greater once marketers develop new applications and eliminate the users' concerns for privacy and security. Until now the marketing applications have been limited to entertainment and weather reports. Further developments in the area of personalization, content management, and infrastructure technologies should enable businesses to make mobile-commerce a part of the customer relationship management process (described in Chapter 3), particularly for time-sensitive applications such as travel.[17]

In addition to the specific examples cited above, there are a number of general trends that are worth observing. Although countries with high income per capita tend to spend more on information and communication technologies, there seems to be a wide spread on the amount spent on information and communication technologies even within countries with similar purchasing power. For example, it is reported that Colombia, Brazil, Romania, and Russia have similar purchasing power. Yet, Columbia spends about seven times as much on information and communication technologies per person as Romania, and Brazil spends about five times more than Russia.[18]

According to one survey, for example, although there is a big gap in the use of information and communication technologies between the developed countries and the developing countries, the gap is narrowing. Rich countries are reported to have 15 percent of the world's population, 80 percent of the world's personal computers, and about 90 percent of the Internet users. However, the personal computer ownership in developing countries is growing at twice the rate of rich countries.[19] Similar trends exist within individual countries. For example, within the United States, Internet access and usage among Hispanics and African Americans is increasing at a much higher rate than for the rest of the population.

As mentioned above, it is important to monitor cultural differences with respect to the adoption of emerging technology so that appropriate product and marketing strategies can be developed to meet and, if possible, exceed the expectations of different segments of the marketplace. The analysis should begin with as basic technologies as websites. Since websites are inherently international, it is important to examine factors that influence their usability. A cultural difference is one important factor. And according to some experts the creation of truly effective websites for international audiences requires carefully designed interfaces rather than simply an accurate translation of the text. Examples of other attributes that could affect usability by international audiences include numbers, currency, date, and time formats, images, symbols and icons, colors and metaphors, flow, and functionality.[20]

As this chapter indicates, airlines will be affected by the changing industry dynamics on the one hand and the changing customer dynamics on the other hand. See Figure 1.8. Three key forces changing the industry dynamics are globalization, declining yields, and segmentation. On the other side, three forces changing the customer dynamics are potential traffic growth related to emerging markets, migration patterns, and the changing and rising expectations of customers.

Technology has the potential to enable airlines to adapt to the changing dynamics of the industry and the customer. In this role, technology is both a driver and an enabler of business strategies. Second, technology is an equalizer of capability enabling small airlines to compete with large airlines, less developed airlines to compete with more developed airlines, and new entrants to compete with incumbent airlines. Consider, for example, the availability of network planning systems over the Web. Traditionally such systems are expensive to purchase and operate. The availability of these systems over the Web reduces their cost of use and enables a small airline to be able to compete effectively with larger airlines not only with respect to having access to sophisticate planning systems but also in compressing its schedule planning cycles. Moreover, application service providers eliminate the internal learning process.

Third, technology has the potential to accelerate the development of emerging markets (relative to the rate of growth of developed markets that are now mature markets). However, the potential explosion of demand from such regions as the Asia-Pacific and Latin America could bring about massive changes in cultural approach and priorities of product development and marketing. Airlines must ensure that technologies fit this radical shift.

In contemporary terms, emerging technology is affecting both the supply chain (channels that link businesses) and the value chain (activities performed by a business to develop, deliver, and support a product). The key to competitive advantage in the changing business environment, as addressed in Chapters 2 through 8, will primarily be in the control of information and knowledge aspects through the careful selection and adoption of emerging technologies.

Challenges Driven by the Changing Airline Customer and Industry Dynamics 25

Changing Industry Dynamics
- Globalization
- Declining Yields
- Segmentation

Airline

Changing Customer Dynamics
- Potential Growth of Emerging Markets
- Migration Patterns
- Changing and Rising Customer Expectations

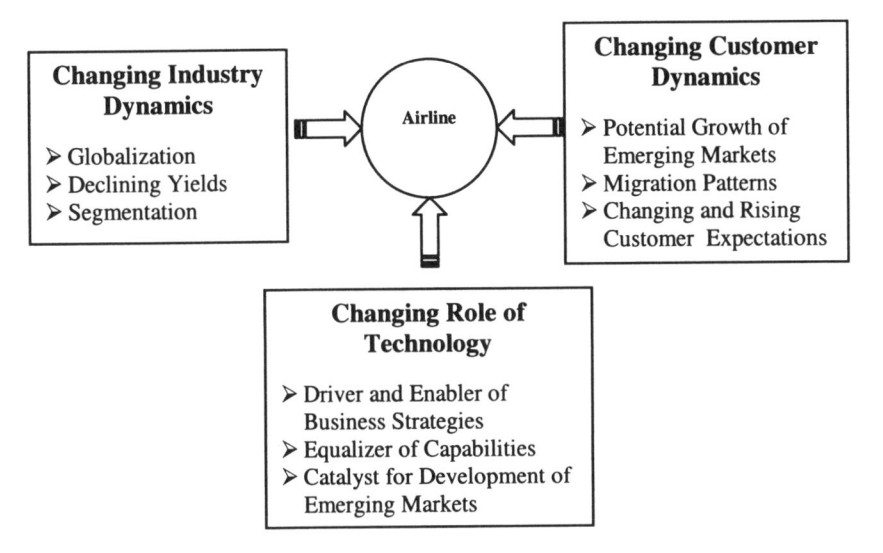

Changing Role of Technology
- Driver and Enabler of Business Strategies
- Equalizer of Capabilities
- Catalyst for Development of Emerging Markets

Figure 1.8 Changing industry and customer dynamics and the role of technology

Notes

[1] Business Week, 3 December 2001, p.43.
[2] Airline Monitor, December 2001, p.11.
[3] The Economist, 30 June 30 2001, pp.51-2.
[4] The Economist, 30 June 2001, p.52.
[5] Boyett, Joseph H. and Jimmi T. Boyett, The Guru Guide to the Knowledge Economy. (New York: John Wiley& Sons, Inc., 2001), p.44.
[6] Pine, Joseph B. and James H. Gilmore, The Experience Economy: Work is Theater an Every Business a Stage. (Boston: Harvard Business School Press, 1999).
[7] Schwartz, Peter, Peter Leyden and Joel Hyatt, The Long Boom: A Vision for the Coming Age of Prosperity. (Reading, MA: Perseus Books, 1999), pp.19-20.
[8] Schwartz, Peter, Peter Leyden and Joel Hyatt, The Long Boom: A Vision for the Coming Age of Prosperity. (Reading, MA: Perseus Books, 1999).
[9] Moore's Law (Gordon Moore—a cofounder of Intel) Power of microprocessors and, in turn, speed, and power of personal computers doubles every 18 months.
[10] Metcalfe's Law (Bob Metcalfe—inventor of network technology called Ethernet) The value of a network increases exponentially to its members with an increase in the number of members—increasing returns in economic terms.

[11] Schwartz, Peter, Peter Leyden and Joel Hyatt, The Long Boom: A Vision for the Coming Age of Prosperity. (Reading, MA: Perseus Books, 1999), pp.22-3.
[12] Evans, Philip and Thomas S.Wurster, Blown to Bits: How the New Economics of Information Transforms Strategy. (Boston: Harvard Business School Press, 2000), pp.36-7.
[13] Moschella, David, "For Wireless, U.S. culture drives down its own path," COMPUTERWORLD, 01 February 1999, p.33.
[14] Shaw, Keith, "Digging into cultural wireless issues," Network World, 11 June 2001, p.48.
[15] Clark, Tim, "Open Source: invasion mode," Asia Week (Hong Kong) 22 January 2001.
[16] Evans, Mark, "How Smart Is Your Card?" E-Commerce Strategies Advisor.Zone 31 May 2001, www.advisor.com/Articles.nsf/aid/EVANSM03, pp.1-4.
[17] Cross, Richard, "Understanding a Blurry M-Commerce Landscape," www.crossworld network.com/article...a%20lurry%20M-Commerce%20Landscape, January 2001, pp.1-7.
[18] World Development Indicators, 2001, p.263.
[19] World Development Indicators, 2001, p.264.
[20] Sear, Andrew, Julie A. Jacko, and Erica M. Dubach, "International Aspects of World Wide Web Usability and the Role of High-End Graphical Enhancements," International Journal of Human-Computer Interaction, 12 (2), 2000, pp.241-61.

Chapter 2

Opportunities Driven by Emerging Technology

With each new generation, airlines have capitalized on advances in technology. Aircraft are now significantly improved in terms of speed and onboard comfort, while operating at lower costs with reduced pollution and shorter trip times in long-haul markets.

The industry has also improved its operations through consistent applications of new technology to its infrastructure. In the fifties, automation improved productivity related to labor-intensive activities, and in the sixties and seventies, the industry optimized planning systems for networks, aircraft fleets, schedules, crew, and maintenance. In the eighties, emerging technology began to play an important role in the development of the product, exemplified by the introduction of loyalties programs, a business-class cabin in long-haul international markets and, later, in-flight entertainment and services such as personal entertainment centers and mobile phones. In the late-nineties, airlines began to examine using emerging technology to get closer to the customer, exemplified by the implementation of customer relationship management systems. In addition, airlines have continuously applied technology to improve their analyses of product profitability, exemplified by the development of sophisticated computer simulation models used to analyze the profitability of a segment, a route, a city, a hub, or a fleet type.

Efforts are now underway to use technology to:

1. optimize operations within a constrained infrastructure;
2. develop and improve on-ground products;
3. strengthen customer relationships;
4. hone measurements of profitability.

This chapter highlights the recent developments in three of these four areas along with an introduction to the technologies currently emerging in each area.[1] Customer relationship management is discussed in Chapter 3. In each of these areas new technologies are being adopted to the extent that they enable airlines to compete more effectively in the new environment that is increasingly constrained in its infrastructure, while at the same time becoming even more dynamic and more competitive.

Operational Planning

The operational component of the airline industry has always been very complex due to the large size and widely spread nature of the network. Operating all day every day all year, in spite of drastic variations in weather, within an infrastructure limited in its capacity is complicated enough. Add to that government safety regulations over aircraft, maintenance and crews, a highly unionized workforce, and the public's growing concerns about air and noise pollution, operational planning becomes a logistical labyrinth.

To deal with this complexity, the airline industry since the sixties has relied heavily on the application of sophisticated analytical models for network, fleet, schedule, crew, maintenance, and airport facility planning. Schedule development models, for example, based on operations research, attempt to maximize the profitability of a proposed schedule by taking into consideration a broad spectrum of variables. Those models took into account the market (passenger and cargo traffic volumes, composition of traffic, and competitive schedules), resources (aircraft, crews, and maintenance facilities), and constraints (route authority, airport facilities, and the environment). By the eighties, technology enabled airlines to apply analytical models to revenue management systems. These systems maximized revenue on the basis of price, first by segment and more recently in terms of origin-destination.

Emerging technology is focusing on decision support systems that (1) work on a real-time basis, and (2) optimize on a cross-functional basis. The airline industry deals with a perishable commodity, highly variable demand, and high fixed costs, and it must deliver its product under circumstances beyond management controls, circumstances such as inclement weather, and air traffic control capacity, procedures, and systems. With so many uncertain factors that change so rapidly and that have such significant effects, airlines have to be as flexible as possible. This means their operations need to work on a real-time basis—or as close to real time as possible—and they also need to be able to respond cross-functionally.

Airlines have traditionally set their schedules six months in advance, based on best available forecasts of passenger and cargo traffic, as well as other conditions such as competitive schedules and infrastructure constraints. Now, however, airlines can adjust their schedules to reflect changes in the marketplace almost up to the time of departure. Four areas of emerging technologies are making this possible.

1. Aircraft manufacturers now offer families of aircraft. These families have similar crew requirements, but each aircraft in the family differs in its payload-range capability so that in selected markets, different aircraft from the same family can be swapped to match capacity with demand, up to almost the last day of departure.

2. Technology is now available to build comprehensive real-time flight-operational databases that contain the most current information on passengers, cargo, crews, aircraft, maintenance facilities and people, and airport facilities and services. These databases manipulate that information to optimize operations on a daily basis.

3. Sophisticated revenue management tools coupled with new channels of distribution—for example, online reservations through the use of the Internet—now enable airlines to manage demand on a near real-time basis.

4. It is now possible—at least, theoretically—for an airline to decide on the order in which its inbound aircraft should be cleared for landing at its hubs. However, even though technology is available to optimize the landing pattern from the viewpoint of the airline, the ultimate decision on the landing pattern rests with the air traffic control authorities.

Most airlines have become fairly sophisticated at optimizing activities on functional levels such as aircraft and crew scheduling, maintenance scheduling, sales, pricing, and revenue management. However, because each function optimizes its own activities with respect to its own variables, there is usually very little integration across functions. One reason airlines have been able to manage individually up to now is because some of these functions or processes have been structured, such as schedule planning. The other reason is that some functions, such as maintenance planning, have been able to work somewhat in isolation due to the fact that each airline has kept its own organizational "silos" of people, skills, and data sources.

Functional departments tend to keep a great deal of information about their own departments, but they have little information about how their individual processes contribute to the company's success. In the case of crew scheduling, for example, does the group recognize the percentage of delays and losses of high-margin passengers that result from its decision to minimize crew costs? While it is true that at least in one area (fleet planning), airlines have typically taken into consideration most of the cross-functional aspects, even here, the impacts have normally been considered in piece-meal and sequential terms, rather than in any comprehensive, integrated manner.

As a result, these departments cannot identify relationships with other departments or indicate how a change in one functional area would affect another functional area. Top management, looking at data and reports generated by each department, cannot effectively assimilate all the discrete data into a complete picture, either. Consequently, improvement initiatives in one functional area may be made unwittingly at the expense of another functional area. For example, a mathematical minimization of crew costs may produce situations that increase delays, a particularly undesirable situation when "pacing" flights are delayed due to crews delayed on other flights.

Significant opportunities for revenue enhancement and cost reduction are to be had by optimizing cross-functionally. One analysis shows, for example, that the integration of network planning, revenue management, and pricing functions could provide an approximate 20 to 30 percent improvement in network contribution to profits. Similarly, the integration of planning activities in three areas (scheduling, crew, and maintenance) could provide an approximate 4-8 percent reduction in costs related to these areas.[2] With such potential profitability increases as these, implementing cross-functional processes becomes almost a requirement for the industry's continued success. The following sections describe the need for airlines to become cross-functional in three areas: revenue management, customer management, and scheduling.[3]

Typically, the activities of at least four functional groups affect revenue: (1) network planning, which is responsible for the development of the schedule; (2) pricing, which is responsible for setting fares and fare conditions; (3) revenue management, which is responsible for controlling the number of seats allocated to each type of fare offered in a market;[4] and (4) sales, which is responsible for selling the seats, negotiating contracts, and establishing promotions.[5] Some of these functional groups, specialized as they may be, work as "silos," and tend to rely on individualized performance metrics that lead to sub-optimal results.

One analyst provides two examples of traditional metrics that are inadequate for managing and measuring revenue. First, while standard route profitability reports may highlight routes where an airline is making or losing money, they do not provide information on the revenue improvement potential of a given route. Second, a revenue index typically assigns target revenue based on the experience of the previous year. Consequently, such a metric does not shed any light on the absolute revenue performance potential.[6] As a result, new metrics are needed to help an airline pinpoint areas where revenue performance could be improved under current conditions. In order to generate these new metrics, airlines must implement new processes that cut across organizational boundaries and functions.

The customer segmentation function or process is similarly disintegrated. Route planners view passenger segmentation in terms of their origin and destination traffic volumes, sales departments focus on corporate segments or travel agencies, and promotions that can be developed that cater to these groups, and loyalty programs look at the total miles travelled by a given segment, and so on. Each group looks at a segment or segments according to its own perspective, and when it decides to cater to a given segment in a particular way, it does so with insufficient consideration given to the impact on the entire airline. Instead, departments need to focus on either a broader variable such as shareholder value, or on a component of shareholder value such as meeting and exceeding the needs of a particular segment with respect to its strategic value and competitive position. This will be discussed in the next chapter.

Customer relationship management systems, which are fundamentally derived from customer segmentation processes, require cross-functionality for similar reasons. Implementing a successful CRM system must involve every function that represents a touch point for the customer, and these functions must work in concert in order to maximize the benefits of a CRM system. An airline using CRM therefore needs to change its thinking and promote the cross-functional framework in its totality, namely, in its organizational structure, data systems, total performance management metrics, and the incentives for cross-functional collaboration.

The need for cross-functional optimization is especially true in terms of scheduling. Airlines typically assign aircraft, crews, and maintenance activities using an approach that involves the use of feedback loops, but which is otherwise essentially sequential. For example, when handling irregular events, airlines tend to focus first on getting the airplanes back on schedule and second, getting the crews back on schedule.

While it may appear that airlines tend to consider the impact on passengers as peripheral to their concerns for operations concerns, the real problem is quite often related to the complexity of scheduling crews. Scott Nason of American Airlines explains the situation. Crew scheduling is complicated because there is a mismatch between what the aircraft can do and what the crews are able to do. See Table 2.1. For example, an airplane can typically fly between 10 and 12 hours per day whereas the crew is typically only allowed to fly up to eight hours per day. Similarly, the length of the day for an aircraft can be 14 to 17 hours whereas it is normally less than 13 hours for the crew. The data shown in Table 2.1 is portrayed for illustrative purposes only. The actual data would vary from airplane to airplane and from airline to airline. Consequently, explains Scott Nason, the only way to get "efficient" crew routings is to mix and match flights to piece together crew routings. Long-haul international flights are an exception. The operations of Southwest Airlines would be another exception where the aircraft keep moving all day and 'sleep' all night.

Passengers currently rank delays as one of their top concerns. The indirect costs in the loss of passenger goodwill due to delays only add to the billions of dollars in direct costs that delays cause. It is essential, therefore, that airlines reduce delays with the passenger in mind. Current processes, however, do not allow for an integrated analysis of the many functions within an airline that may be impacted by irregular events, which makes reducing delays that much more difficult to accomplish. Therefore, airlines must adopt a different mindset in order to manage delays.

Table 2.1 Why crew scheduling is so complicated

Activity	What an Aircraft Does (Typical Day)	What a Crewmember is Allowed to Do
Flight Hours	10-12	Less than Eight
Length of Day	14-17	Less than 13 (plus report and debrief)
Length of Overnight	7-12	More than 10 (including report and debrief)
Return Home	Multiple bases (any one every 3-4 days)	One home, each 2-4 days
Delays, Overfly	No problem	Above limits still apply

Source: American Airlines

A different mindset to manage irregular operations is required for two reasons. First, as with many other industries, more and more products and services in the airline industry are produced by a network of organizations—some within the control of an airline and some outside the control of an airline—operations such as catering, ground services, and maintenance. To be able to respond effectively to irregular events, groups need to be managed in concert so that everyone is notified of each other's actions and what is required of their group in particular. Second, airlines simply must become more responsive to customers. The general trend across industry lines is toward greater flexibility in the delivery of products and services to meet customer needs. Airline passengers bring with them the expectation that they will receive similar levels of reliable and flexible service, and they are disappointed if those expectations are not met. Therefore, airlines must shift their primary focus from aircraft and crew to customer. Recall the point discussed in the previous chapter about moving to an experience economy.

Consider, for example, the impact of a 30-minute delay involving an aircraft that has 50 passengers compared with one that has 350 passengers. Standard metrics currently used to measure on-time performance are inadequate to the task of improving that performance. It is important not only to know how to measure it but also the causes of poor on-time performance, such as maintenance, ground services, and crew. Such metrics indicate, among other things, the percentage of flights departing and arriving within a 15-minute deviation of the scheduled time. These measurements do have value for the purpose of making intra-industry and historical comparisons, but they do not reflect the causes of or contributing factors to delays. Without the ability to recognize the cause, the problem cannot be resolved. And because the causes of delays can include a complex relationship among maintenance, ground services, and crew, airlines need to develop new metrics that incorporate a cross-functional approach.

The old Swissair developed just such an approach. The Operations Control Department designed a cross-functional operations management system of metrics focusing on total performance management. The previous Swissair team, led by Daniel Weder and Volker Heitmann, refined the total performance management philosophy that had initially been developed at the University of St. Gallen in Switzerland.[7] The operations group, responsible for any schedule change up to seven days before departure, needed a system that took into consideration the impact on aircraft turn-around of a broad spectrum of services provided by maintenance and ground handling, among others.

The Swissair Operations group thus developed a project, now known as virtuGate that measures total performance based on a set of "Zoom Indicators", so-called for their ability to pinpoint impacts on a detailed level, and a set of "Gates" that represent the interface from the customer process to a supply process. Under this system, every function or process can keep its specialized indicators to measure its own performance, but the process of each function must also be integrated into the common framework of the airline. This common framework is characterized by a pre-identified, limited set of Zoom Indicators that express the airline's vision. Consequently, all employees work toward maximizing their performance on the basis of those shared indicators. They are able to share the same vision while they are also aware of their individual contribution to the total airline.

Looking at the functionality of just one Zoom Indicator gives a basic understanding of how the virtuGate system works. One Zoom Indicator relating to delays is Passenger-hours (Pax.h), or passenger hours wasted, from the beginning to the end of the value chain, on such activities as waiting to check in, waiting for a delayed flight, or waiting for delayed baggage. Each player in the value chain's contribution to Pax.h is coded separately, so that the initiator of the delay can be identified. Pax.h can also be expressed in terms of dollars by assigning a monetary value to an hour of a passenger's time. Pax.h not only replaces a number of other traditional, individualized metrics, it also enables the airline to identify the processes causing the delay. With this highly flexible and relational information, management of groups at both the cause and the effect sides of the delay can identify, evaluate and recommend the investments needed to eliminate or reduce delays.

Figure 2.1 represents a typical virtuGate display of a sample delay scenario. This photo is difficult to read since it is taken directly from a computer screen where the original screens are larger in size and the information is displayed and coded in color. Nevertheless, it is illustrative of the of the type of information, the sources of the problem, and the costs and benefits of alternative solutions.

The *y* axes of the graphs display the Zoom Indicators for Pax.h, CF+ (positive cash flow) and CF- (negative cash flow). The colors of each segment within a bar on the graph represent the various supply processes that affect Passenger Hours. The colored buttons at the bottom of the display are similarly coded to each group and represent the links to the

Gates (supply processes) of the next layer (maintenance, air traffic control, and ground services). CF- shows the negative impact on processes due to deviations from plans (for example, lost revenues due to flight cancellations). CF+ shows the positive impact on processes due to deviations from plans (for example, additional revenues generated by the

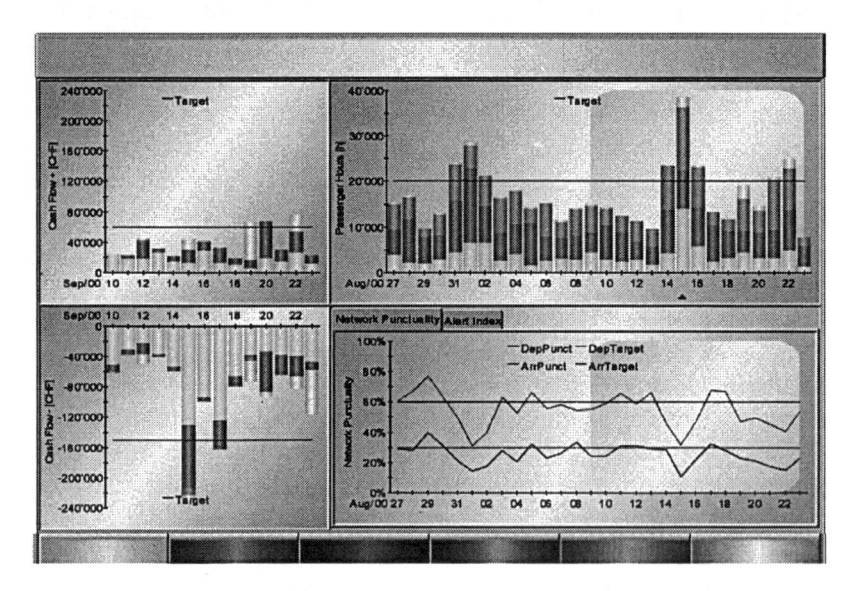

Figure 2.1 virtuGate passenger hours and impacts on cash flow
Source: Swissair and virtuGate

use of a large aircraft). Each of these indicators is also time-linked, so that the indicators can be linked to each other in time, and so that they can ultimately be tracked to related time-specific events such as the occurrence of a snowstorm.

This display allows management to identify, at a glance, where each group needs to improve and where each group contributes to the success of the airline. Cross-functionality, which is the true nature of any airline's operations, also becomes immediately apparent in very detailed, yet actionable terms. Implementing this kind of understanding of cross-functionality beyond the scope of delays, beyond scheduling, to the total airline will similarly reveal true causes and effects within the airline as a whole.

The use of such integrated metrics may indicate surprising results. An airline may discover that it is not in fact cost-effective for it to commit the necessary resources to achieve the best on-time performance in the business. It may actually be more cost-effective to be the second-best, given the cost of requisite additional airplanes, crews, maintenance facilities, inventory, and airport facilities. A truly cross-functional system can take into account all of these factors, as well as block times and ground times, handling of passengers, baggage and cargo, and hub structure variables such as number of gates and gate assignments. External factors could also be included, such as ATC and airport procedures. Finally, systems are now available to address 'robust scheduling' in which a system proposes a series of cross-functional actions, for example, to recover time lost due to operational disruptions at a minimum cost to the whole organization.

In general, the broader the scope of an airline's cross-functional vision, the more likely the airline will be to incorporate those functions in a total system management program, and the easier it will be for an airline to maximize performance within each function and throughout the total airline. Such a move will fit well within the context of the experience economy mentioned in the previous chapter. The 11 September events bring into greater focus the value of cross-functionality as additional security requirements create a new challenge for the on-time performance of an airline. Now, an airline must balance not only passenger convenience with its operational requirements but also its own security requirements as well as the security requirements dictated by governments.

Product Development

Aircraft

Airlines have traditionally excelled at applying technology to the aircraft itself. In particular, airlines have devoted a lot of attention to the configuration of the cabin. Airlines have gone from a one-class cabin, to two classes, to three classes, some back to two classes, and then back to three classes. British Airways recently introduced even a fourth class cabin that provides more space than the standard economy-class seat but is available at lower costs than the cost of a business-class seat. For the cost-conscious business traveler and the economy passenger who is willing to pay a little more for the extra space, the seats in this section are wider and with more legroom. Although this reduces the overall capacity of the airline, it exemplifies British Airways' strategy to concentrate on the high-yield segment of the travel market.[8] There is also a continuous debate on whether airlines should offer first-class service in trans-Atlantic and trans-Pacific long-haul international markets. Major airlines such as American, British Airways, and Lufthansa believe that the first class product is necessary and an essential element of their full-service. Although it may be true that a significant portion of the first-class cabin may in fact be used by passengers who have been upgraded, that may well be part of the reason why first class exists at all.

According to one analysis, despite the extraordinary differential between first class and economy class fares on intercontinental routes, incremental costs outweigh the incremental revenue. See Figure 2.2. The analysis begins with the cost of an economy-class seat and adds the incremental costs of additional space, cabin crew, catering and amenities, and seat configuration. The configuration effect makes adjustments for spreading the fixed costs (pilots, maintenance, and so on) over fewer seats. This provides the number for the cost per seat for the first-class product (sixth bar from the top). Next, the analysis makes an adjustment for the lower load factor in the first-class cabin to provide a figure for the cost per passenger (second bar from the bottom). Comparing this cost per passenger to the revenue per passenger (bottom bar), it is clear that the first-class product is not cost effective. Results would obviously vary by aircraft, by route, and by aircraft type.[9]

Figure 2.3 shows a similar analysis performed on the economics of the business-class product. The superior profit margin of the business-class product is striking from this analysis which also substantiates the strong interest by most carriers in such a product. Finally, Figure 2.4 shows an analysis of an even more innovative and profitable product—a premium class economy seat. It was invented by Virgin Atlantic and followed, many years later, by British Airways. Even though this product is not anywhere near as highly contested as the business-class product, it appears to be even more profitable than the business-class product.[10]

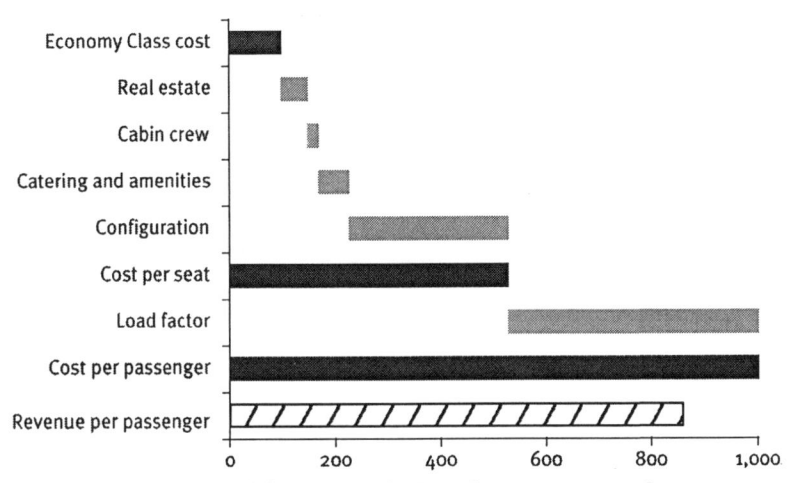

Figure 2.2 Economics of first class indexed to economy class
Source: Andrew Lobbenberg, Flemings Equity Research

Despite the higher profit margin of the premium economy class product, most airlines are continuing to focus on further innovations in the premium class product, for example, seats that convert into flat beds. This innovation may be so important to some passengers—especially those taking long-haul flights—that they may choose a particular airline based on this one element. Other onboard features being introduced include specially designed swivel seats that allow up to four passengers sitting in the center of the first-class cabin to have face-to-face meetings, a folding

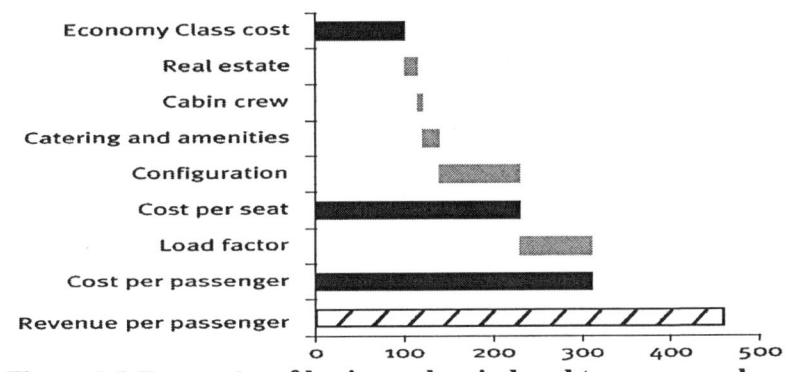

Figure 2.3 Economics of business class indexed to economy class
Source: Andrew Lobbenberg, Flemings Equity Research

40 *Driving Airline Business Strategies through Emerging Technology*

Category	Value
Economy Class cost	~100
Real estate	
Cabin crew	
Catering and amenities	
Configuration	
Cost per seat	~150
Load factor	
Cost per passenger	~150
Revenue per passenger	~150

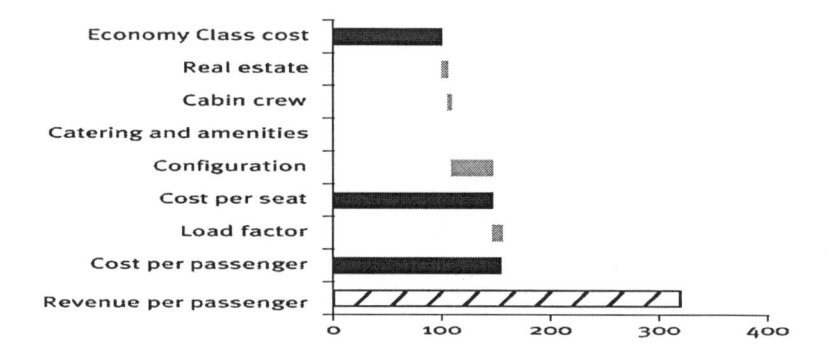

Figure 2.4 Economics of premium economy indexed to economy class
Source: Andrew Lobbenberg, Flemings Equity Research

table that can be used for a business meeting or dining arrangement for two, beauty therapists, an exercise bar at the back of the cabin for simple stretching exercises, and lavatories in business class that are 50 percent larger and include windows. As discussed in Chapter 5, the Airbus A380 is one aircraft that will provide new opportunities for product development by increasing the degree of customization to match the changing and diverging needs of each segment of the marketplace.

The gradual trend toward ever-longer trips is encouraging airlines to examine cabin configurations that provide passenger areas to sleep, stretch their legs, or alleviate the monotony. Fortunately, emerging aircraft technology with, for example, a full stand-up height lower deck provides wide array of potential product innovations for all classes of service. To begin with, the lower-deck area can be used to provide real beds. See Figure 2.5. The lower-deck area can also be used to offer a casual lounge, a bar, or duty-free shopping. Other possibilities include locating some lavatories in the lower deck. These lavatories could then be made larger and additional washrooms located in adjacent areas. Relocation of lavatories to the lower deck provides advantages: (1) space for extra seats on the main deck; (2) removal of noise and bustle surrounding the lavatory areas; and (3) an opportunity for passengers to stretch their legs even more.

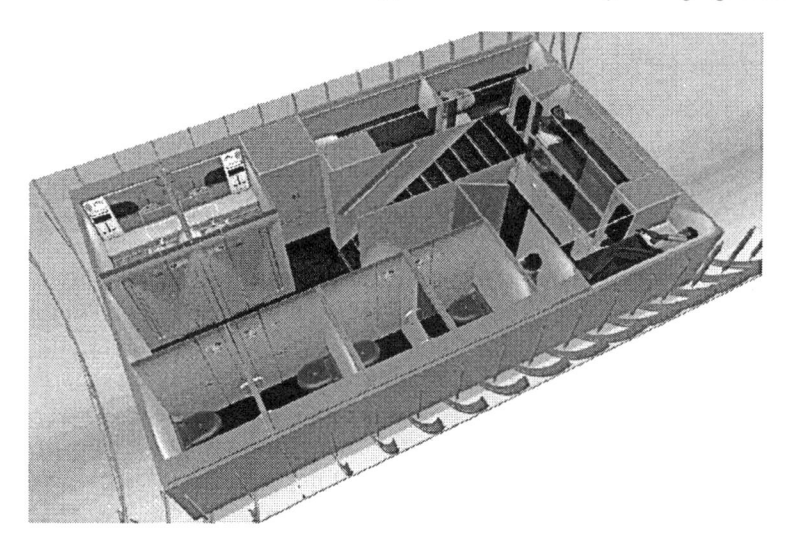

Figure 2.5 Potential product innovation on ultra long-haul sectors
Source: AIRBUS

Airlines have also introduced an array of in-flight entertainment and services. They can now offer a full library of in-flight movies using Digital Versatile Disk technology, and in-flight satellite television is not far behind. Airlines are also focusing on providing Internet access and e-mail capability. Some passengers want the ability while on board to conduct all aspects of e-commerce such as shopping, making a broad spectrum of reservations, and, in general, surfing the Web; others just want the ability to send and receive e-mail. At the present time, in-flight access to the Internet can be made possible by connecting a laptop to an on-board server that is connected to a satellite that can relate the information to the ground. This service currently is not as optimal as it could be in that the air-to-ground connection is made through a narrowband link, which significantly hinders transmission of data that typically requires broadband connections. To upgrade the technology to match passenger expectations for broadband capabilities, airlines would have to make significant investments. Those investments, however, could translate into a potential source of revenue for airlines.

On the Ground

While airlines have concentrated so much of their efforts on enhancing service in the aircraft, they have focused less on innovations on the ground. This is partly due to the fact that airlines have less control over ground features. For example, airports control the number and location of gates and the quality and quantity of passenger facilities and services, while governments dictate airport security, immigration and customs, capacity of the air traffic control system, and the capacity and the condition of the airport access and egress system. This does not mean, however, that airlines cannot affect service on the ground, and in fact, airlines are making some enhancements in this area.

To respond to passenger frustrations with slow airport processing procedures, more and more airlines are installing self-service machines such as ticket kiosks that can be used by frequent flyer card-holders or by e-ticket holders with standard credit cards. These machines allow passengers to check-in, select seats, receive boarding passes, and receive baggage tags, all without having to wait in long lines for service.

As airlines are working to increase Internet accessibility in the air, they are also increasing Internet access on the ground, particularly through wireless capabilities. Such services will enable passengers in airport lounges to use their laptop computers to access the Internet through wireless connections. While such service provides added value to the customer and also presents additional means of generating revenue, three problems are associated with establishing wireless capabilities. The first problem is the result of the amount of wireless traffic within the airport environment. Second, since different airlines support different Wireless Local Area Network (WLAN) Internet Service Providers (ISPs), there is no assurance that all passengers will have access to their service providers throughout an airport. Consequently, it may become necessary for airports to adopt a common infrastructure that allows access to all WLAN ISPs. Third, the use of wireless devices on the ground by a potentially great number of passengers and service staff can create conflicts among devices used in critical airport communications and navigation. Continuing developments in wireless technology may resolve some of these problems.

Speech recognition is another emerging technology that can enhance service to customers on the ground. It can provide voice message responses to customer inquiries relating to flight status information, reservation information, information about gates and misplaced bags, and the status of requested upgrades. One advantage of such systems is that they can be available on a continual basis so that passengers do not need to be placed on hold, and they can handle high volumes of inquiries that especially occur during operational difficulties caused by bad weather. It is also possible that speech recognition could be used for a multitude of languages. One caution relevant to this application is that translations into other languages need to be accurate not just linguistically but also culturally, and those translations still need to communicate the brand images of each airline.[11]

The ultimate on-ground innovation is being conceptualized by the Simplifying Passenger Travel (SPT) project that is led by the International Air Transport Association. The object of this project is to streamline repetitive checks of passengers and their documents by collecting necessary information once and then sharing it electronically with all the service providers at an airport in the passenger's travel chain. The innovation would rely on the use of an internationally compatible multi-functional device, such as a passenger card with microchips that can be coded with passenger-specific information. Along with this device, airports would use biometrics (machine recognition of physical characteristics such as a passenger's iris or fingerprints) to identify passengers, and international agreements to abide by common standards and practices. The concept's goal is to provide a hassle-free travel experience—and after the 11 September incidents a safe travel experience—for the passenger and a more efficient use of resources by all service providers involved in processing a passenger through an airport.

Figure 2.6 illustrates the "one-stop" check-in process envisioned by the SPT concept. It is assumed that a passenger, call her Jane, makes a flight reservation using her travel device and a personal computer. She connects to a dedicated website to make her reservation, and relevant information about her reservation is stored on her travel device which would also contain other pertinent information about Jane's preferences for seats and meals, and whether she is a frequent flyer. Her reservation could be made according to those preferences. Once the reservation is made, Jane would automatically be informed of any special requirements for travel to her destination such as the need for a visa. She is then prompted to pay for the ticket using a credit card or a direct transfer from her bank account to the airline's bank account. She would also be informed that if she wished to check her bags before arriving at the airport, she could do so at a designated off-airport, secure baggage check-in facility. This facility would also require Jane to use her travel device.

Once at the airport, Jane would proceed to the one-stop gate and insert into a reader her travel device that is encoded with her flight information. If the travel device does not have that information, Jane can type in the information about her destination and the name of the airline and then work through an interactive screen. Once the reservation is located, Jane is welcomed by name and asked to complete a biometrics scan. Biometric information can also be contained in the travel device or it can be stored in a centralized database(s). If Jane's identity is verified through a successful match between the biometrics scan and the information contained in the travel device, the one-stop system provides information on Jane's boarding gates and seat assignment. If the identity is not verified, she is asked to see a customer service agent.

If she has not used an off-site baggage facility, Jane can check her bags at this point. She would place her bags on an automated device that registers the number of pieces and their weight and produces baggage tags containing radio frequency (RF) chips that can be used to track Jane's bags throughout the handling system. The RF chip can be embedded in the baggage tag. However, as the RF chips are relatively expensive at the present time, they can also be built into suitcases. This alternative enables the chips to be reused. The system can also assess any excess baggage charge that Jane can pay for with a credit card. Jane can attach the tags herself, or an airline staff member who is expected to be monitoring multiple check-in stations can also assist. The system would also print boarding cards for all segments. With those boarding cards, Jane could then proceed to a security checkpoint or exit passport control if required by the control authorities.

While Jane proceeds to the gate, the one-stop check-in system relays ahead Jane's relevant information to both the embarkation airport passport control authority and to the control authority at the final destination airport. The authorities at the destination airport can then reply with an 'OK to board' message which avoids the admissible passenger problems that exist today. A similar procedure is already in use with the Australian electronic visa system. Control authorities at either point of departure or point of arrival can request an interaction with Jane, either through a series of simple questions on a screen, or in person.

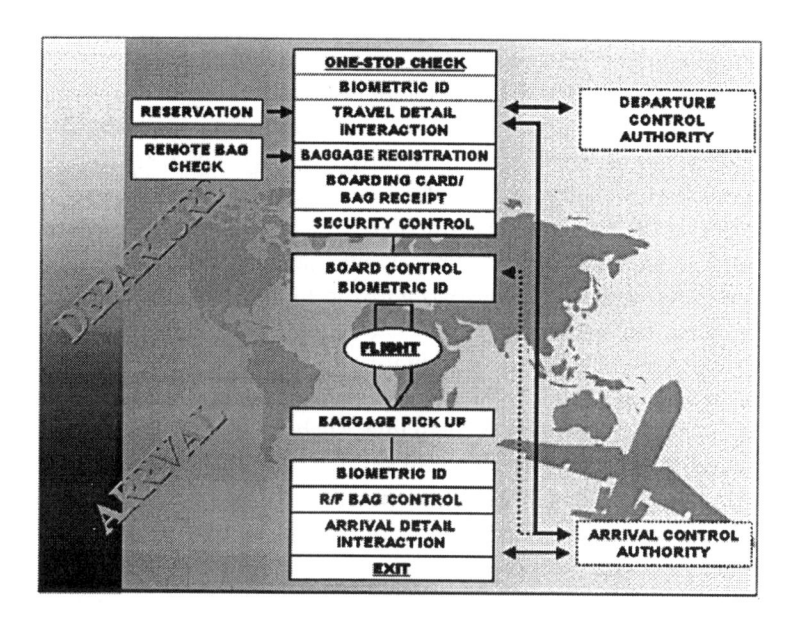

Figure 2.6 Simplifying passenger travel project process flow
Source: The International Air Transport Association

At the departure lounge, Jane would use her travel device, go through another biometrics scan, and assuming a successful match, be allowed to board the aircraft. This second biometric check at the gate would be purely an identification check and reconciliation of baggage. The system would confirm that the baggage checked by this passenger matches the passenger who has boarded. If a passenger checks baggage and does not board the aircraft, the system would issue an alert and the baggage would be removed from the aircraft. The second check is important in that if additional questions and checks are required, they need to be conducted prior to boarding the aircraft.

At the destination airport, Jane would pick up her baggage in the designated area and would then be allowed to exit the gate using her travel device again, unless there is a discrepancy relating to the bags. The system would ensure a match between the passenger and the bags by making use of the travel device, the biometrics scan, and the encoded baggage tags. For international travel, unless the control authorities wish to interact with a passenger, the destination gate would open to allow the passenger to exit the secure area and enter the arrival area.

This concept is still very much in the development stages. Even if it never comes to fruition in exactly this way, the importance of the SPT concept is that it represents the kind of large-scale vision required to streamline the airline industry, especially uneven and sometimes lengthy delays as a result of the 11 September incidents. For such vast procedural change to come about, individual airlines would need to coordinate their procedures internally, and the airlines would need to agree on standardized procedures and data management.

Other concepts related to facilitating passenger and baggage handling are nearer at hand. In regions where airport capacity is constrained and good rail service is available, such as in Europe, airlines are seriously beginning to look at using rail service in short-haul markets to feed their long-haul markets. Such cooperation between airlines and railroads would not only free airport capacity but also provide passengers with faster access between airport and city center. Although the use of intermodal transportation is not new to the aviation industry, its implementation has been challenging. Coordinating the transfer and security of baggage has been difficult, and real-time information on the status of flights and gates along with train schedules has not been available, and ticketing for combination air and rail travel has been inadequate. Technology has advanced in all these areas such that airlines are now testing intermodal linked services at major European airports such as Frankfurt and Paris' Charles de Gaulle.

Another means of facilitating travel is the envisioned introduction of large business jets dedicated to trans-Atlantic operations. These jets could provide nonstop service from regional airports in Europe to major airports in North America or conversely from regional airports in North America to major airports in Europe. Eventually, service could also be provided between regional airports on the continents. Aircraft that would be potentially capable of such flights include the Gulfstream V, Bombardier's Global Express, the Airbus A319 or the Airbus Corporate jetliner, and the Boeing Business Jet. The availability of smaller airplanes with transoceanic range can now provide potential opportunities for all business class scheduled operations in secondary markets. Consider, for example, the availability of Airbus 319 and Boeing Business Jets. Both have the potential to offer service in all business class configurations at fares lower than the standard business class fares in such trans-Atlantic markets as London's Stansted Airport to New York's Kennedy Airport. Smaller aircraft such as the Gulfstream V or Bombardier Global Express could be scheduled for even smaller markets such as London's Stansted to New York's White Plains or Westchester County Airports. Such a product would be appealing to price-conscious business passengers who travel in point-to-point markets. It would be particularly attractive to business travelers who do not work for large corporations that have the clout to negotiate large corporate discounts with mainline network carriers. In addition to a potential lower fare, some passengers could look at such service to be of higher quality in the sense that the airport processing time would be less and passengers would be getting a 'private business jet.'

Although there is a significant unexploited opportunity in large markets, airlines—new or incumbent—were reluctant to develop such a product in the past due to the high cost per seat and/or incompatible payload capability of past aircraft. The large capacity of the aircraft would mean low frequency that would in turn reduce the benefit of point-to-point service. Business passengers need high frequency to accommodate their need for flexibility as well as potential delays and cancellations due to mechanical reasons. Insufficient demand in secondary markets makes the availability of high frequency uneconomical using large aircraft. The frequency hurdle, coupled with the mainline carriers' frequent flyer loyalty programs whose value increases with the size of the network, deterred new airlines from offering such products. However, the large mainline carriers were reluctant to offer such products, fearing traffic diversion from their own services in the adjacent markets.

Now the situation has changed. New technologies, new business models, and new experience are changing the landscape. Not only do aircraft such as the Airbus Corporate Jetliners (ACJs) and the Boeing Business Jets (BBJs) have desirable trans-Atlantic capability—exemplified by the operations of a A319 ACJ aircraft for the Daimler Chrysler Corporation between Stuttgart, Germany and Pontiac, Michigan—but they also have good economics. While such aircraft have so far been deployed in business class configurations, there could be a significant market if the product was based on the current super economy class such as the one introduced by British Airways. Such a product has the potential to provide higher profit margin on a unit of space used than the standard business class product across the Atlantic. Such products could be introduced by totally new airlines or by existing airlines as part of their new products. If such a service is offered by existing airlines, the product would not be that much different than the Concorde service provided by Air France and British Airways, both of whom benefited from its "halo" effect.[12]

Challenges

Although emerging technology provides tremendous opportunities for product innovation, its implementation raises a number of issues and challenges. First, there is the question of the rate of return on investment. Not only is the initial investment relatively high for some of these technologies, but product longevity is also becoming shorter. This is in part due to changing customer expectations and partly due to the speed with which new technologies are developed. Consider as an example an airline that invests huge amounts of money in new business-class seats with, say, a 55-inch pitch, 150-degree reclining capability, and individual in-flight entertainment systems. A year later a competitor introduces seats with 60-inch pitch and a laptop power port built into the seat. Another year later, another competitor introduces a flat-bed seat that takes the same amount of room as traditional seating. Then still another competitor introduces Internet access at each seat. At what point does the airline further invest and change its product?

One possible solution to this problem might be to implement a modular approach by which an airline could add features at a regular pace. However, such an approach would require manufacturers to offer flexible aircraft interiors, and in-flight entertainment and seat vendors to develop standardized platforms that are adaptable.[13] But it is also possible that an airline could establish workable contracts to this effect so that it could thus adapt to the changing demands of a dynamic market.

Relating to new passenger processing concepts like the SPT project, in order for such expansive visions to be realized, all parties involved in the processing of passengers need to adopt common standards. Those standards would apply to the chip architecture embedded in hand-held multi-functional devices and in baggage tags, and to the biometric passenger identification systems. In the ideal case, all airlines need to adopt common processes, procedures, and systems, just as all airports need to adopt universal systems.

Facilitating such common standards quickly raises the question of who controls the airport infrastructure. Is it the airline, the airport, or a government agency? Who develops the required databases, who pays for them, and who manages them? Currently, airports around the world have different standards and procedures for security, immigration, and customs control. Are these airports and government authorities willing to change? Could all international airports agree to pre-clear arriving passengers on the basis of information provided by different airlines? Assuming that agreements acceptable to all could be achieved, would all the airports and airlines be able to fund such technological developments? Currently, most of the high-technology systems described above are being developed to meet the needs of frequent fliers. Should these systems also be available to infrequent travelers? Implementing and maintaining two parallel systems would present enormous logistical difficulties. But would it be cost-effective to implement such high-end technologies for all passengers, regardless of their profitability?

Such questions ultimately relate to the degree of customization airlines will provide. In general, airlines need to determine to what degree they can personalize their services. In actuality, customizing every aspect of air travel simply is not feasible. Consider, for example, a scenario in which an airline allows each of its passengers to pre-select their meals and drinks and the times that they are to be delivered. Such a service alone would create numerous logistical problems for the cabin crew, already working with limited resources and often under severe time constraints.

While these challenges are very real, there are solutions. Regarding the development of common standards and architecture for the SPT project, the goal is not to produce a solution but rather a framework that would enable different airlines and different airports to develop their own solutions to deploy common protocols. The IATA is one of 11 organizations on the SPT Board. The SPT interest group itself includes several major airlines, airports, customs and immigration authorities, and technology companies. The potential for the SPT group to produce a workable framework is therefore high, given the strength and composition of the interest group, and the fact that a framework would allow for realistic and functional variations.

Regardless of how airlines meet the particular challenges of implementing new technologies, they must above all ensure that any innovations they adopt are meaningful. Passengers must first of all be willing to accept any new features—an issue of particular concern to the potential use of biometrics—and passengers' concerns about the privacy and security of data must be respected. Any new features must also be user-friendly. Information provided to passengers through websites, for example, must be available as close to real time as possible and must be easy to access. And finally, product innovations have to provide recognizable, tangible and intangible benefits for which the passenger is willing to pay. In essence, if a customer cannot use a product, does not like it, or does not see the value in it, the product offers no value to the airline, either. Once again, this aspect calls attention to the experience economy.

Airline Profitability Measurements

The value to measure accurately the profitability of the airline at various levels is so vital that it can in fact be the sustainable competitive advantage as good as if not better than the usual low cost, product leadership, or operational excellence drivers of competitive advantage. On the negative side, the accurate measurements of profitability have become increasingly difficult due to proliferation of hub-and-spoke systems and strategic alliances as well as a much broader spectrum of products and customers, on the one hand, and the lack of consistent, robust, and timely information, on the other hand. On the positive side, emerging technology has enabled the development of sophisticated analytical systems. Such systems, coupled with the use of integrated and timely databases and data warehousing and data mining techniques can support decision making by providing vital information on the profitability of individual products, segments of customers, or parts of a network. This section provides an overview of emerging directions in the measurement of airline profitability based on the works of a number of airline analysts.

The structural difficulty relating to the measurement of product profitability within the airline industry relates basically to two aspects as explained by Ben Baldanza of US Airways.[14] First, flight segments— which are a function of fleet and schedule structure—drive costs whereas fares are primarily a function of a passenger's points of origin and destination. The issue relates, then, not so much to the lack of accuracy regarding the absolute measurement of costs and revenues but much more to the alternative and often inconsistent decisions relating to either the allocations of costs to an origin-destination basis or fares to a segment basis. Second, in this industry, capacity-related costs are much higher than traffic-related costs, encouraging management to maximize the revenue earned from selling the fixed schedule. The impact of these two factors is becoming increasingly complex due to the proliferation of liberalization policies and the resulting hub-and-spoke systems as well as a broad spectrum of fares and fare instability situations.

The value of network management becomes clearly evident from the analysis of a former American Airlines executive, Tom Cook, who examined the contribution to airline revenue of two major management areas, network management and revenue management. Cook estimated that through a careful management of its network an airline has the potential to improve its total revenue by about six percent. The majority of airlines have been able to achieve an improvement between one and four percent. And even the leading airlines have only been able to achieve up to five percent improvement in their revenues. Consequently, there is still an unrealized potential of one percent through a better management of networks. Similarly, Cook points out that revenue management can improve total revenue by as much as eight percent and, here again, even the leading airlines have an unrealized potential of about one percent.[15] Recall the earlier discussion that cross-functionality can enhance profit margins even more.

Let us first consider the emerging developments relating to the management of networks. Ben Baldanza of US Airways provides a number of key insights in this area. First, technology is now available to measure profitability at a disaggregate level of the network, for example, a segment, or an origin-destination market. Second, for each segment flown, it is now possible first to measure incremental revenue as well as incremental costs, and second to measure the financial contribution of each segment to the overall network. Third, sophisticated techniques are now available to measure revenue and costs allocations. For example, airlines need sophisticated revenue proration methodologies in the ever-changing and complex environment encompassing enormous reliance on hub-and-spoke systems and strategic alliances.

With respect to costs, there is a need to take into account such costs as the costs of marginal capacity, and fully allocated costs. Examine for a moment the information shown in Figure 2.7 that contains hypothetical data for illustrative purposes.[16] This airline is transporting a substantial

52 *Driving Airline Business Strategies through Emerging Technology*

number of passengers whose fares are covering just the marginal costs and not even the capacity costs, let alone the fully allocated costs. Marginal costs are short-term costs and fully-allocated costs are long-term costs. Therefore, in the case of this particular airline, many passengers are paying less than the long-term cost of carrying them.

In addition, there is even a greater need to take into consideration the real costs of network connectivity. Finally, Baldanza points out the most difficult part of the cost analyses—the need to measure network opportunity costs. For example, airlines quickly take into consideration the revenue contribution of a connecting passenger. They assume that the seats available on a connecting flight would have gone empty. However, this assumption involves an opportunity cost as there is always a probability, no matter how small, that some of those seats may not have gone out empty.[17]

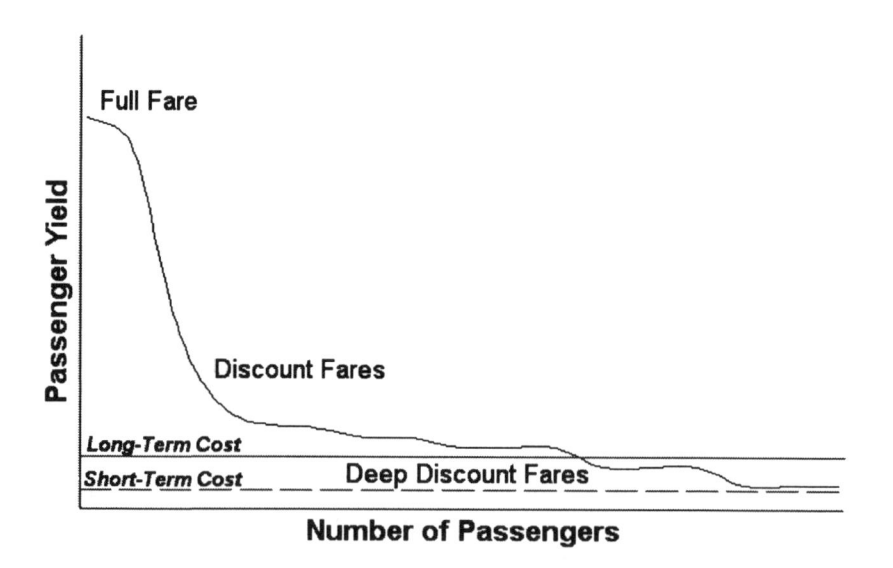

Figure 2.7 Variation in the profitability of economy-class passengers

One major airline, for example, is measuring the profitability of various types of point-to-point and connecting passengers, disaggregated by length of haul and type of fare. See Figure 2.8. Although this figure contains illustrative data, it does show that this airline is losing significant

Opportunities Driven by Emerging Technology 53

money on two types of passengers: (1) passengers traveling on economy fares in short-haul point-to-point markets; and (2) passengers traveling on economy fares and making a connection between short-haul and long-haul markets. Conversely, passengers traveling on premium fares in either long haul or short-haul point-to-point markets provide high profit margins. The data shown in Figure 2.8 would indicate that this airline should reduce its capacity that is being sold to passengers traveling in short-haul point-to-point markets at economy fares and those making connections between short-haul and long-haul markets. Alternatively, the airline needs to examine if it is possible to serve these passengers at a lower cost. Second, this airline needs to examine if it is possible to carry more passengers in the long-haul point-to-point markets and try to sell these seats in the higher-fare buckets. Finally, this airline needs to defend and grow the long-haul and short-haul point-to-point segments traveling at premium fares.

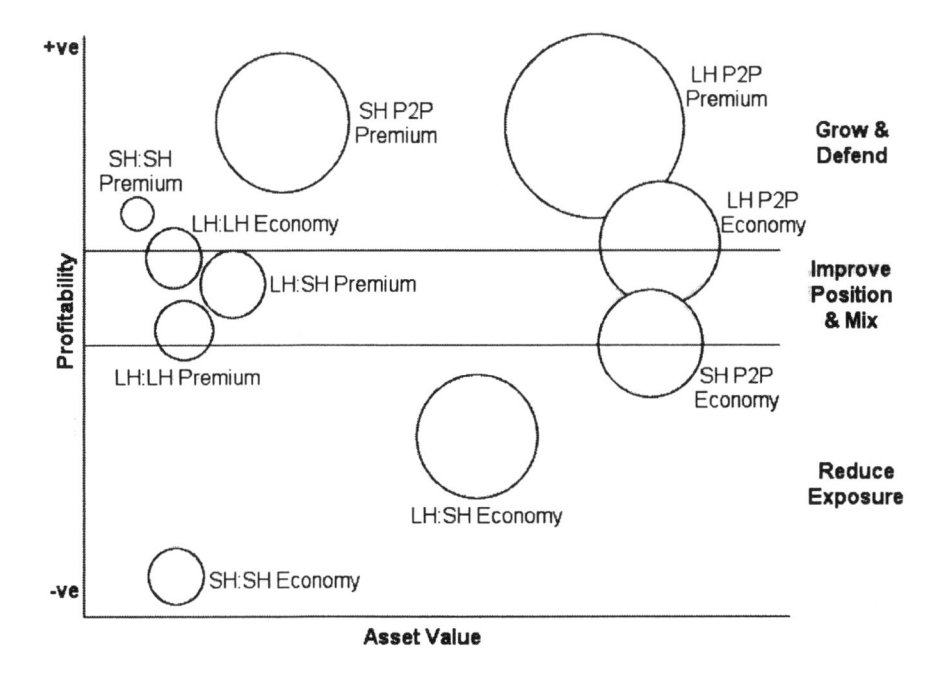

Figure 2.8 Segmentation profitability
Source: Based on Aviation Strategy, Issue 42, April 2001, p.7

Other analysts provide additional valuable insights into the emerging directions for route and network profitability. Consider, for example, the potential distortion by taking into consideration book value of assets such as maintenance facilities and aircraft. The use of book value of maintenance facilities could distort the "true" economic profitability of flights if maintenance costs incorporated in the analysis are low due to the fact that the finance department has "written off" the cost of maintenance facilities. It is necessary to use market-rate value to obtain a more realistic assessment of flight profitability. Similarly, it would make more sense to use the replacement value of aircraft rather than actual value. The use of market-rate values provides a decision-oriented view of flight and route profitability analyses.

Airlines have generally been at the forefront of adopting emerging technology to reduce costs, improve revenue, or improve the service provided to customers. This chapter has highlighted the contribution of emerging technology with respect to operational efficiency, product leadership, and product profitability. However, to fully capitalize on the value of emerging technology, management needs to agree on some basic principles relating to profitability analysis. First, it is difficult to envision the implementation of sophisticated technologies and techniques when the airline does not even have a control over the data definitions, that is, a common language and understanding of key measures. What is a passenger? What is a flight? Second, what financial performance metric should the airline use? Should management use profit margins based on long-term or short-term basis? Should the airline maximize revenue per available seat kilometer *generated* or revenue per seat kilometer *sold*. Third, which costs are fixed and which costs are variable? Should an airline take into consideration the costs of customization of aircraft? An agreement on the basic principles and data definitions will enhance substantially the value of emerging technology.

Notes

[1] The rationale for the first three areas is based on Michael Treacy and Fred Wiersema, *Discipline of Market Leaders* (Reading, MA: Addison-Wesley, 1995).
[2] Kabbani, Nadar, "Key IT trends and opportunities in the airline industry," IATA Information Management Conference presentation, Amsterdam, The Netherlands, 24 April 2001.

[3] These three areas are derived from a total of six areas described by Lucio Pompeo in "The cross-functional airline challenge," *Aviation Strategy Management,* July/August 2001, pp. 15-18.
[4] As well as controlling numbers of seats by fare type and market, revenue management now also tends to adjudicate on discounting or commission deals.
[5] Pompeo, Lucio, "Who is responsible for revenue?" *Aviation Strategy Management,* March 2001, pp.17-19.
[6] Ibid., p.17.
[7] Weder, Daniel, and Volker Heitmann, "The Next Generation of Strategic Controlling," IATA Information Management Conference presentation, Amsterdam, The Netherlands, 24 April 2001.
[8] Shifrin, Carole, "Atlantic Luxury," *Airline Business,* June 2001, p.53.
[9] Lobbenberg, Andrew and Stephen Clapham, "Beds and Bollinger," A Global Report on Airlines, Flemings Research, 12 April 2000, p.6.
[10] Ibid., pp.9-10.
[11] Canday, Henry, "Speech prompts," *airline info tech,* November-December 2000, pp.22-3.
[12] Piling, Mark, "Power broker," *Airline Business,* June 2001, p.98.
[13] Piling, Mark, "Flights of fancy," *Airline Business,* January 2001, p.47.
[14] Baldanza, Ben, "Measuring Airline Profitability," Butler, Gail and Martin Keller (eds), *Handbook of Airline Finance* (New York: McGraw-Hill, 1999), pp.147-59.
[15] Cook, Tom, "Creating Competitive Advantage in the New Millennium Using Decision Support Systems," IATA Information Management Conference presentation, San Jose, California, 12 April 2000.
[16] *Airline Fleet & Asset Management,* Issue 13, May-June, 2001, p.61.
[17] Baldanza, op cit, p.156.

Chapter 3

Market Segmentation and Customer Relationship Management

As stated earlier one of the major forces affecting the business world is the degree to which customers are influencing or even dictating new business practices, essentially taking control from suppliers.[1] It is no surprise that leading companies are analyzing their markets to identify the most valuable customers as well as the appropriate strategies to use in developing relationships with these customers. For financial institutions and telecommunication companies such strategies have included developing a one-to-one relationship with customers through the use of market segmentation and customer relationship management (CRM). This chapter addresses both of these strategies and their application to the airline industry.

Prior to 11 September, many airlines were examining the implementation of CRM as part of their long-term strategic positioning in their pursuit of premium passenger traffic. However, after 11 September, some airlines lost interest in their CRM initiatives, to reduce budgets and/or to focus on attracting the lost traffic. These decisions may not necessarily have been cost effective as the CRM initiative can, in fact, add value in both areas. In the short run, these initiatives can help an airline target better its offers using the available data. In the longer run, an airline can redefine its touch points not only to provide higher-quality service and make travel simpler but also safer.

Market Segmentation

Segmentation is the process of separating customers (actual or potential) into groups according to common characteristics so that marketing and operational strategies can be targeted at specific populations.

Segmentation in the airline business has typically focused on defining business travelers versus leisure travelers for the purpose of developing schedules and pricing policies. In addition to identifying a passenger's purpose for flying, the airline industry has also used a number of other segmentation criteria, such as socio-demographic characteristics, status in the frequent flyer programs, and more recently, distribution channels. Table 3.1 provides examples of airline-oriented market segmentation criteria and individual samples of each criteria point.

Evolving Strategies

Originally, airlines segmented their customers based on demographics. The basic problem with this strategy was that significant differences among individuals could exist within the same demographic group. For example, all individuals identified as Baby Boomers are not likely to have the same preferences and buying behavior. An airline therefore cannot identify the right product for the right customer, even though customers have been defined demographically. Consequently, segmentation based on demographic characteristics alone is not consistently actionable.

Airlines then further segmented the market based on socio-economic characteristics within demographic groups. However, even this step had limited value in that an airline could still make assumptions that are not necessarily true. An airline might assume that passengers with low incomes are likely only to be interested in the lowest-fare seats, and segments with high incomes are more likely to be interested only in premium-fare seats. In reality, high-income groups may be interested only in low-fare seats if they do not perceive an additional value in high-priced seats.

Pairing socio-economics with demographics still was insufficient to identify customers accurately. A high-income Baby Boomer, assumed to prefer premium-fare seats, might still prefer lower fare seats, or a low-income Generation Xer might not in fact have a flexible schedule, and his or her ticket might actually be paid for by a higher income relative. Therefore, more customer data was essential to predict an airline customer's buying trends.

Table 3.1 Market segmentation criteria for airlines

Criteria	Sample Identifier
Initial	
Demographics	Baby Boomer
Socio-economics	High-income
Advanced	
Trip purpose	Business travel
Frequent flyer status flights	Miles earned through company airline
Distribution channel	Tickets purchased through a travel agent
Emerging	
Needs	Purchase of discount fares with flexible scheduling
Behavior	Heavy use of long-haul flights
Perceived value	Better food on board is worth the higher-priced ticket
Customer valuation	Lifetime value (revenue less costs), range of influence, growth potential

Some airlines now segment the market based on five additional criteria: trip purpose, frequent flyer status, club membership, first-class upgrades, and distribution channel. Each of these criteria provides more information, but even these elements are insufficient.

Segmenting customers according to trip purpose has usually meant identifying two major categories of travel, business and leisure. Most airlines have done a reasonable job of developing marketing strategies based on trip purpose. Prices, seat inventory, and fare restrictions have been used effectively to maximize the total revenue from each flight. Knowing that a customer is traveling for business reasons, for example, has enabled airlines to charge as much as five or six times the average price paid by leisure passengers traveling on the same flight with the same class of service.

Segmenting by frequent flyer status focuses primarily on usage, that is, how many miles a passenger flies. While there is likely some relationship between usage and loyalty and value, it is difficult to define how that level of usage—and perhaps loyalty—translates into profitability. Although most airlines can provide a list of their frequent flyers ranked by the amount of travel, very few can provide a similar list ranked by some measure of profitability. Indeed, some airlines admit that some percentage of the top-tier frequent flyers may in fact be providing very little profit, if any. If a frequent flyer purchased most of his or her tickets in the low-fare buckets by using special corporate discounts (or used expensive redemptions, for example, upgrades with some special displacement costs), the airline would not be realizing much profit from that customer.

Consider the information shown in Figure 3.1, illustrative as a typical large airline. An analysis may reveal, for example, that 44 percent of the passengers in its frequent flyer program are considered to be unprofitable, 39 percent have a low value, 13 percent are valuable, and only four percent have high value. Within each of the four categories, there are passengers who belong to the top-tier frequent flyer status and others who are regular members. In the unprofitable category (44 percent of the total passengers in the frequent flyer program), 97 percent are regular members and three percent have top-tier status. Top-tier members in this category are passengers who have been able to find ways to take advantage of various discounted fares. This number is significant since it is a small percentage (three) of a large number (44). The airline is obviously over investing in this group of frequent flyers.

At the other extreme are four percent of total frequent flyers who are in the category of very valuable customers. Eighty percent of this group has top-tier status and 20 percent are regular members. Here the airline is likely to be under investing in these high-value customers. Because this 20 percent of the passengers do not fly very frequently and therefore do not belong in the top-tier status, they do not receive the same privileges received by the top-tier status group. Nevertheless, because of the high fares they pay, they generate high levels of profits. In summary, out of every 1000 passengers in its frequent flyer program, this hypothetical airline appears to be over investing in about 40 passengers (a combination of unprofitable and low-value groups) and under investing in about 33 passengers (a combination of valuable and high valuable categories).

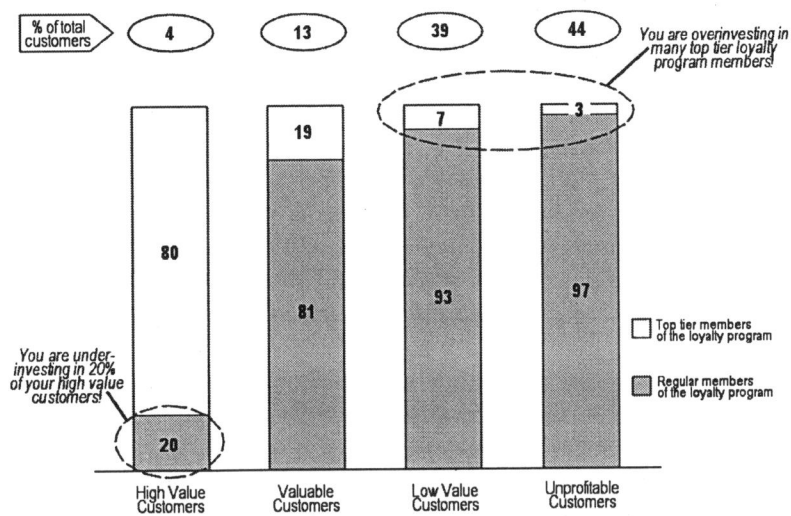

Figure 3.1 A Value-based segmentation relating to loyalty program status
Source: Roland Berger

Currently, many airlines reward frequent flyers based strictly on their absolute mileage status in the program (status vs. bonus miles). But the total miles a passenger has traveled does not yield much information about a passenger. Were the miles accumulated very rapidly or over a long period of time? Did the passenger earn the miles by making non-travel related purchases or by actually flying the airline? Even if the miles were accumulated through travel, did the passenger fly this particular airline or one of its strategic partners? Passengers traveling on the airline, paying premium prices, and traveling often represent a greater value to the company. Being able to identify which of its frequent flyers fall into this higher value category would be extremely beneficial to any airline.

Much the same is true in regard to distribution channel, which refers to the means by which customers purchase their tickets. Did a given passenger purchase his or her ticket through a travel agent, or from an Internet company? While collecting this kind of information has enabled airlines to launch some targeted marketing initiatives, segmenting passengers according to distribution channel has had only limited value. Like all the other segmentation so far described, distribution channel does not allow an airline to determine the profitability derived from the customers within that group. The most accurate means of determining profitability comes from assessing each customer's needs.

While the trends in segmentation up to now would lead one to segment customers on finer and finer levels of data within a category, such an approach does not solve the fundamental problem of airline market data. For example, determining which customers who use the Internet to make reservations also use the airline's website as opposed to another online organization such as Travelocity, Expedia, Priceline, or Orbitz indicates nothing about whether a customer is satisfied with any one of those services. Similarly, further segmentation of leisure travel by type of leisure (for example, vacation versus visiting friends and relatives) adds only limited value. The key to the effective marketing of an airline's services is to obtain a comprehensive understanding of the true needs and preferences of different segments of passengers. Knowing a customer's needs allows an airline to anticipate a customer's choices. Developing strategies and products based on needs means that when a customer looks for the exact service to fit his or her needs, the airline would already have that service in place for the customer to purchase it. In this way, the airline can attract and retain those segments of the market it has identified as the most profitable.

Individual customer needs are often a complex combination of several different kinds of specific criteria. For example, passengers buying their tickets through Priceline are clearly very price-sensitive but they are also very scheduling flexible with respect to such considerations as choice of airline, routing, number of stops, and time and days of travel. Focusing on a customer's multiple needs—price-sensitivity and flexibility in scheduling—will help an airline market its services to meet each of those needs, rather than just one and not another.

Segmentation Techniques

In order to build a database of customer segmentation criteria, an airline must first determine what segmentation technique would best collect the data. There is no single technique that is best for all airlines or even best for a given airline in every situation. In general, an airline should identify its customers with respect to their needs, service expectations, and attitudes to products and services.

The Travel Research Center (TRC) in Australia, which has done extensive research in segmentation based on work done by Focus Research in New Zealand, has analyzed the value of six segmentation criteria: geographic, demographic, behavioral, psychographic, personality, and needs. These segmentations are just examples (for both multiple business travelers and single leisure travelers) and they do not cover all behaviors in an individual market.

The geographic criterion can apply to segments based in certain locations, for instance, business travelers from the U.S. West Coast. Demographics refer to criteria such as gender or age. The behavioral criterion refers to usage and consumption, for example, the frequency with which a passenger flies or the heavy use of long-haul flights. Psychographics refers to values, attitudes, and lifestyles. Personality criteria are indicators of a traveler's demeanor, such as easy going, or demanding, or wanting to be in control. Needs refers to what attributes the passenger is looking for from the brand.

The TRC then crosses the segmentation criteria with a number of considerations such as generalized versus targeted markets and unimodal versus multimodal markets. Generalized markets refer to the whole market while targeted markets refer to a specific group such as a particular brand, product, or usage. Unimodal markets refers to the fulfillment of one need such as safety, while multimodal markets refers to the fulfillment of multiple needs such as safety as well as convenience and competitive fares.

Once the criteria are crossed with the considerations, the customers' needs within each category become more apparent. Those needs can then be generalized from a tabular form into customer identity, attitudes, and emotions. Figure 3.2, illustrates the application of such an approach for an airline such as Southwest Airlines. In this example, a passenger's functional needs may be to arrive at the destination in the shortest time without paying too much money, while still enjoying the

64 *Driving Airline Business Strategies through Emerging Technology*

flight. From the airline's viewpoint, marketing initiatives that may satisfy these needs may be point-to-point service, low fares, no frills, and good service. The same figure also illustrates social values or branding personality an airline could develop in response to its customers' identity needs, and the overall company symbols that could speak to the emotive needs of the airline's customer base. In the original exhibit, the text appeared in different colors to correspond to the colors of the ovals.

Get to destination
Shortest time
Not pay too much
Enjoy the flight

Functional Needs
Identity Needs
Emotive Needs
Symbolism
Social Values
Product Features

Smart
No nonsense
'Just an ordinary guy'

Modest
Belonging

Honest
Straightforward

Egalitarian/Inclusive
Value for money
Down to Earth

Point to point
Low fares
No frills
Good service

Illustrative

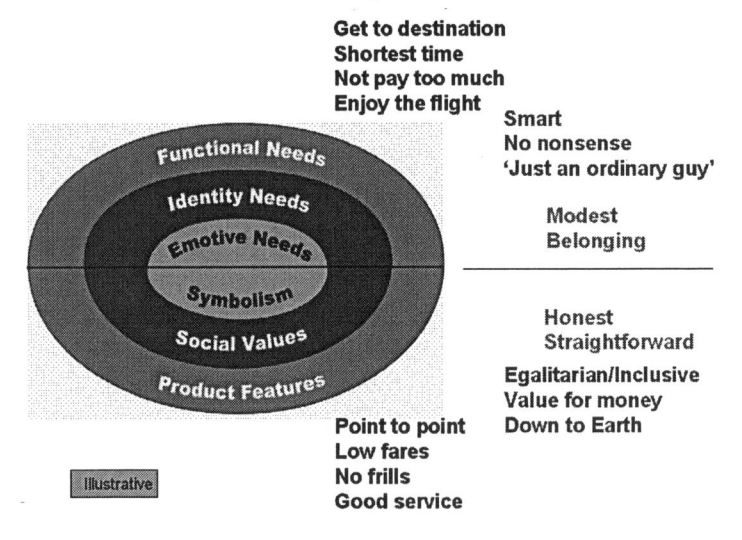

Figure 3.2 Product development based on need brand relationship: Southwest Airlines
Source: Taylor Nelson Sofres and Focus Research

A second example of an emerging segmentation approach also comes from the work of the TRC in Australia. In this approach, statements made by business travelers are clustered to identify potential segments. Table 3.2 shows some examples of such statements. Based on an analysis of such statements, TRC was able to group these statements into, for example, five clusters. In this case each cluster was titled National Flag Flyer, Switcher, Pragmatist, Risk Averse, or Mileage Mad. See Table 3.3. Generalizations could then be made about passengers who might be grouped into one of the five clusters.

These approaches to segmentation can be very useful for identifying, developing, and implementing profitable marketing initiatives. For example, let us take the cluster, "switcher." One option could be to stop investing in this cluster of passengers. Another option could be to convert these passengers into another cluster such as "Mileage Mad." Similarly, statements shown in Table 3.4 made by leisure travelers can provide valuable insights for strategic and tactical marketing initiatives.

Table 3.2 Examples of business traveler statements

Traveler Statement
I generally like to fly the same airline—I know what to expect
I look for airlines that treat me like an individual, not a number
I usually fly the same airline to build frequent flyer points
I choose airlines that recognize me with upgrades and special treatment
I select airlines that provide good in-flight service and products
I often fly the airline of my destination because they know the territory better
I tend to change airlines to sample new products and services
I don't mind which airline I fly, they are all the same
I feel happier flying my national airline
I still get excited about flying
I think flying is a chore; all the glamour is gone
I need value for my money, so I look for airlines with low fares or deals
I choose successful or profitable airlines to fly with
I enjoy flying most airlines; it broadens my experience
I just want the airline that can get me there quickest; nothing else matters

Source: Travel Research Center, Australia

Table 3.3 Examples of clusters based on the TRC analyses

Cluster Type	Passenger Description	Potential Type of Airline
National Flag Flyer	Passengers likely to fly the national flag carrier	National Flag Carriers like British Airways or Singapore Airlines
Switcher	Passengers who are not brand loyal and are likely to be attracted by the prospect of a new experience	Start-up airlines or airlines that promise unusual on-board comforts or airlines that fly to "exotic" places
Pragmatist	Passengers concerned about costs, schedules, or frequent flyer issues	Airlines that offer low fares or discounts and strong network or schedule
Risk Averse	Passengers who try to avoid risks and do not want to make mistakes	Airlines that are known and familiar and those with an excellent safety record
Mileage Mad	Passengers who are focused on accruing miles	Airlines that offer the highest return for the lowest number of miles, and airlines with extensive networks or a large number of partners

Source: Travel Research Center, Australia

Market Segmentation and Customer Relationship Management 67

Table 3.4 Examples of leisure traveler statements

I fly:
    To relax
    To try something new
    To feel excited by new experiences
    To experience a culture different from my own
    To meet interesting and friendly local people
    To have good food and wine
    To do lots of different things
    To meet new people
    To experience an exotic atmosphere
    To learn something new
    To go somewhere hot and get some sun
    To reward myself for hard work
    To revisit a well-loved place
    To get away from crowds
    To do something few others have done

Source: Leisure Travel Monitor, Travel Research Center, Australia

From the above statements, The Leisure Travel Monitor of the Travel Research Center of Australia deduced the insights displayed in Table 3.5.

Table 3.5 Insights from leisure traveler statements

These travelers tend to be young, polarized on income
They usually travel with a partner (or occasionally alone)
They just want to relax and get away from it all
Although they tend to plan ahead, they are not avid planners or information collectors
They tend to make reservations long before the flight
They tend to use major chain travel agents
They tend to take relatively long holidays (usually 2-3 weeks) that represent tailor-made trips
They tend to stay at middle- to upper-tier hotels

Source: Adapted from Carolyn Childs presentation at the Ohio State University's Eighth International Aviation Symposium, Porto, Portugal, 31 May 2001

Again, this is one example of a segment that happens to be significant. Other airlines can perform similar analyses by first surveying customers for their reasons for traveling and then performing a statistically-driven cluster analysis.

The key to effective marketing of an airline's service is to develop a comprehensive understanding of the needs of passengers within each segment. One analyst recommends, for example, the division of business segments into three needs-based sub-segments:[2]

1. Business travel decisions controlled by strict and enforced corporate travel guidelines regarding choice of the airline and fare;
2. Business travel decisions based on corporate guidelines but with consideration given to efficiency of the choice regarding airline and fare;
3. Business travel decisions left completely at the discretion of the traveler.

Similarly, the leisure market can be divided into sub-segments based on socio-demographic characteristics such as age, income, marital status, and willingness to pay a certain fare. In both cases, for business and leisure sub-segments, the airline needs to create, communicate, and deliver unique value. For a segment of passengers with families, for example, the airline could promote stress-free travel for passengers with children. The airline could offer lower fares for families and on-board children's meals, or they could also provide special family-focused entertainment and service facilities at airports. To propagate the success of such a program, any time a family took advantage of such an offer, the airline could collect information about the number and ages of the children so that current data could be kept about them and their changing needs as they mature.

Although sub-segments based on needs provide more detailed information about customer groups, those groups still must be analyzed before an airline can develop the appropriate strategies for each group. The ten bases on which to evaluate each group are: size, expected growth, profit margin, degree of competition, robustness, predictiveness, measurability, projectability, branding considerations, and the gap between perception and reality. Attributes of expected growth, profit margin, degree of competition, measurability, and branding considerations are relatively straightforward, but the others deserve some explanation.

The size of the segment can be measured with respect to number of passengers, dollar value, or potential market share for the airline. Regardless of the type of measurement employed, the important attribute relating to segment size is value, not volume.

Robustness relates to the degree to which observed customer behavior will remain constant over time. Passengers' pattern of preferences can and do change with changes in life, and robustness helps to quantify the degree to which a given need may change over the lifetime of a customer. It is important to determine the constancy of an attribute because value propositions based on non-robust segments may lead an airline to produce and deliver more of what it is already good at rather than products based on the expected, albeit changing, needs of customers.

Predictiveness is a very important attribute and it can help answer the following: Can the segments distinguish meaningfully among airlines or product brands? Can actions by an airline be effective in influencing the behavior of passengers in these groups?

Projectability relates to an airline's ability to understand how it is meeting and could meet a segment's current and future needs. Take, for example, a particular sub-segment within the larger segment of business travelers that flies in business class across the North Atlantic. This segment could prefer to choose an airline that provides flat beds in business class and meals available in a special airport lounge before boarding long-haul, nighttime flights. Or this segment might soon prefer business-class service in a smaller, standard-body aircraft that flies non-stop between two smaller cities. If an airline projects that this North Atlantic business class sub-segment will move toward this service trend, the airline can evaluate its position in that market and its current and future competitors in that market and thus determine if it wants to offer that kind of service in the future.

The gap between perception and reality refers to the difference between what potential or even current customers believe about an airline versus what is actually true. Those beliefs apply to both tangible and intangible characteristics. Some airlines based in developing countries complain, for example, that despite the fact that they have a very good product in the market, they cannot attract certain segments. This may result from the fact that passengers may not be familiar with the airline and therefore may simply wish to avoid the potential risk of poor service. Or, some people may not even call a full-service airline on the assumption that its fares would be higher than a low-fare airline. Some people may not fly on an airline based in a third-world country based on the assumption that it is not safe. Some passengers may not be willing to buy a higher-fare ticket thinking that all passengers receive the same service.

The gap between perception and reality is a difficult attribute to measure, but is an important one to track. Unless an airline knows what passengers believe about its services, it is unable to develop marketing strategies to combat those beliefs. Once known, that information may influence branding or marketing strategies across the entire airline's products and services.

It is important to keep in mind that many of these attributes are not independent of each other. Consider, for example, size and growth. A small segment may be more important if it is growing very rapidly than one that is larger but is experiencing negative growth. Similarly, a small segment may be more valuable if it has a much higher contribution per passenger than a large segment. Likewise, a small segment may be more valuable if the airline has a larger market share of it than a segment in which it has a much smaller market share. Some segments (such as risk averse) may be highly desirable but difficult to influence. For example, according to the TRC, foreign overseas airlines are currently more popular among younger travelers in Japan. However, it is the older, more established travelers who represent higher value. But these passengers are difficult to acquire and retain. Consequently, should Western airlines focus on easier but less profitable passengers or tougher but more lucrative passengers? In the final analysis, it is the dollar value—and the return on dollars—that is of the greatest concern.

The key ingredients for segmentation are, first, collecting the raw data itself. Most airlines have developed their passenger databases using some very basic information on their frequent fliers, information that does not sufficiently reflect buying and usage behavior. Such information can be collected by a combination of surveys and actually tracking behavior during various phases of the trip. For example, when using the online booking channel, did the passenger ask for the lowest fare? At the airport, did the passenger use the airport lounge and specific services such as check-in, bar, or the Internet? During the long-haul flight, did the passenger sleep, work on the laptop, or watch movies? The means of collecting these types of data and the privacy issues associated with it are discussed briefly in the next chapter. Suffice to say for now that airlines could conceivably collect a great deal meaningful information about their passengers than they currently do.

One key criterion for airline customer segmentation is whether an airline is pursuing identifiable or non-identifiable customers. An identifiable customer is one where an airline can link a unique customer to his or her own data. This requirement allows an airline to place a passenger as an individual in a certain segment based on the data that the airline can clearly attribute to the passenger (for example, past flights of the passenger). A non-identifiable customer is one predicted to have a certain flight activity profile based on certain demographic and socio-economic characteristics. Airlines, with their current loyalty programs, should actually be able to segment identifiable customers. This issue of identification is critical for the implementation of effective CRM. For example, if an airline does not know if a particular passenger used the airport lounge, how can the airline target the passenger to promote the usage of the lounge and its facilities? Most airlines only keep statistics of the total number of passengers using the lounge and then infer from this data the percentage of the top-tier members which uses the lounge regularly (non-identifiable). There is no way of knowing, from this data, if a particular member ever used the lounge.

Once the data is collected, it then needs to be analyzed especially in terms of the costs and profitability associated with passengers whose needs are fully identified. Customer profitability can be calculated by comparing the profit margin within a given segment and then cross-referencing each of the margins across each segment. For example, what are the airline's true costs of providing service to those passengers who have bought their tickets from an Internet company like Priceline versus the cost of service to passengers who bought their tickets direct from the airline? Identifying those true costs means identifying costs also associated with related factors. What was the average reduction in empty seats on the flight? Did higher-paying passengers experience reduced comfort with respect to such aspects as longer check-in times, overcrowding in gate areas or on the airplane itself, and less storage space in overhead bins?

It is possible that the costs of reduced passenger comfort might outweigh the benefits of those empty seats being filled by passengers who pay significantly less for their fares. But because the airline industry does not yet collect needs-based information, we don't really know the amount of potential dilution resulting from the lack of ability to satisfy all the walkup business. Until all these factors can be determined and the costs and profitability of each evaluated, an airline will not be able to develop truly effective marketing strategies.

Customer Relationship Management

Chapter 1 discussed the intensification of competition in the airline industry, resulting in a continuous decline in passenger yield. During normal recessions—and special incidents such as the Gulf War and the September 11 attacks—yield and premium traffic decline. As a result, airlines are always searching for ways to maintain and increase their share of premium traffic and invest their marketing dollars intelligently. The judicious use of Customer Relationship Management (CRM)—to build win-win, customer-centric strategies—may well be a cost effective way for an airline to compete in the current and emerging landscape.

What CRM is and What it is not

CRM is a management paradigm that has the potential of (a) converting a production-driven airline into a customer-driven airline, and (b) raising significantly an airline's efficiency and effectiveness. For example, by integrating numerous databases, CRM can provide personalized service and targeted offers based on passenger profiles and shift certain passenger segments toward the lower cost distribution channels. CRM is about acquiring customers, developing them, retaining them, and thus maximizing the profitability of various product lines. One of the basic principles of CRM is that the size of the ultimate segmentation process is *one*, meaning that particular relationships can and should be developed with each individual customer. While this level of segmentation may be impractical within the airline industry, effective segmentation strategies coupled with an integrated CRM infrastructure can increase revenue, decrease costs, build and retain a loyal customer base, and attract new, profitable customers.

Figure 3.3 provides one analyst's view of four major components of CRM: (1) customer valuation; (2) touch point logic; (3) service personalization; and (4) campaign management. The first component shows the need for and the role of a customer value model that focuses on

Market Segmentation and Customer Relationship Management 73

current and future value of a customer as well as underutilized potential value. This component of the CRM process calls for continuous mapping of customer needs as well as drivers of customer purchasing decision. The second decision relates to the touch point logic to structure CRM measures. Three examples of touch points are the website, the reservations center, and the airport processing facilities. The third component deals with treatment management to personalize the service provided to a customer. This component has three sub-components: (a) definition of treatments; (b) access to individual customer information; and (c) the decision criteria regarding who receives which treatment. The final component—campaign management—calls for the establishment of an automated information structure and cultural change to transmit targeted offers to targeted customers.[3]

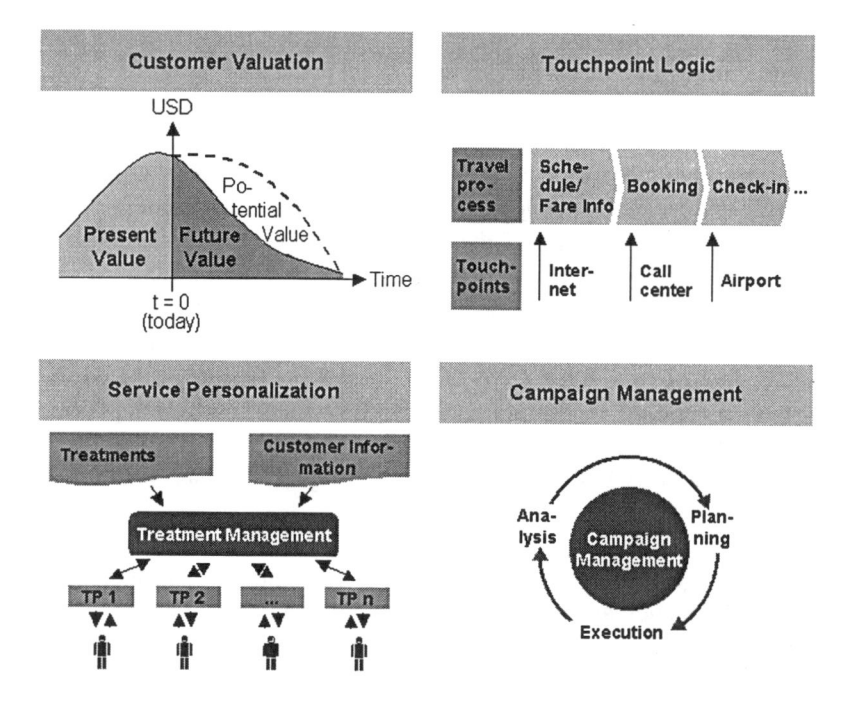

Figure 3.3 Four major components of a CRM initiative
Source: Roland Berger

Having briefly described what CRM is, it may also be useful to remark on what is not CRM. First, it is not a "plug-and-play" type of system that an airline can buy from a technology vendor. Second, it is not a system for focusing and winning any customer. It calls for investing in high-value customers or at least finding ways to serve low-value customers differently, that is, with much lower costs. Examples include directing low-value customers to automated call-center features (instead of live customer service agents) and less compensation for denied boarding, lost baggage, and lower priority service in cases of missed connections. Third, it is not a data collection, manipulation, or retrieval system. It only requires the use of a comprehensive database. Fourth, it is neither inexpensive nor easily adaptable to existing legacy systems.

In the final analysis, CRM is a management paradigm to build a customer-centric organization by combining (1) various information technology systems and tools such as the Internet and wireless communications, (2) a broad spectrum of business processes such as customer segmentation and valuation as well as customer service and care, and (3) employee training and incentives. This CRM management paradigm allows an airline to provide different passengers with different levels of service, recognition, convenience, and special privileges for their loyalty. For airlines, some of these rewards up to now have consisted of special check-in counters at airports, upgrades to a higher class of service, preferred seat assignments, opportunity to board the airplane early, and access to special toll-free telephone numbers for reservations. More recently, special privileges have taken the form of timely and interactive information relating to the status of flights, check in by telephone, and relevant special offers.

Although CRM is not a new subject, it has become more prominent in recent years with:

1. The emergence of enabling technology (the Internet and technological progress in such areas as data warehousing and data mining);
2. Customers' increasing expectations and choices;
3. The desire to seek new ways of developing sustainable competitive advantage;
4. Airline product convergence and aggravated competitive pressure.

The primary reason for the recent increase in the use of CRM is the development of the enabling technology that makes easier to implement it on a more widespread basis. With these newer informatics technologies, airlines can track their customers on several data points, such as profitability, behavior, and level of satisfaction. This information is stored in a database that airlines can then mine to build more cost-effective relationships with those customers. It is essential that the database be accurate, expandable, robust, and "mineable."[4] In the context of CRM, term data mining refers to the use of analytical tools and techniques to analyze the customer data to (1) identify insightful patterns, and (2) interpret these patterns to design meaningful and timely marketing programs.

Finally, it is important to point out that there is no unique way to implement the CRM initiative. Figure 3.4 shows two alternative methods that have been used by two different airlines. One airline started by segmenting its customers by value. This airline then examined different treatments that could be provided to different customer segments. A second airline first examined different treatments that could reasonably be implemented. This second airline then identified the customer segments who should be receiving each type of treatment identified. The first method tends to be slow but is targeted and measurable. The second method tends to be fast, comprehensive and pervasive. However, it is more difficult to measure results.

Critical Success Factors

Many companies have made considerable investments in establishing a CRM program only to realize limited benefits. To ensure that a CRM system is truly successful, an airline should focus on each of seven steps crucial to the development of a profitable CRM initiative.

The first factor critical to the success of CRM is where it is positioned within the airline. This decision is crucial because positioning will have a direct effect on how the program is implemented, which will in turn dictate its ultimate success. The IT department at first seems like the logical place to position CRM because the IT staff would be the most adept at implementing or acquiring CRM systems. But marketing or sales departments would be the primary users of CRM, as would operations departments. However, the choice of where to position CRM is not as simple as choosing among IT and operations and marketing.

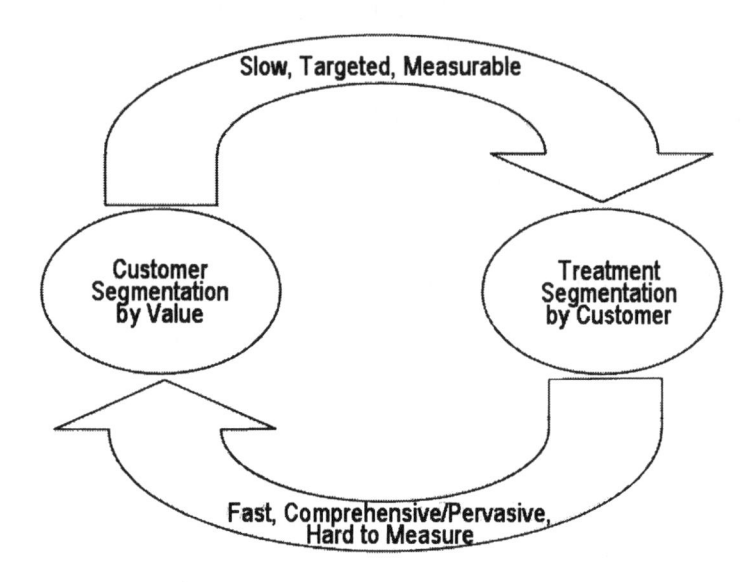

Figure 3.4 Alternative methods for implementing a CRM initiative

An integrated CRM initiative captures data across the spectrum of an airline's activities—from flight reservations to baggage claim, and even beyond to activities relating to customer retention. Information also comes from different entities such as individual passengers, corporate customers, travel agencies, and company staff. Capturing information from all customer touch points and making that information widely accessible within the airline is an absolute necessity for the success of CRM. Therefore, CRM needs to be integrated throughout the airline's corporate structure.

Accomplishing such a company-wide integration presents particular challenges for airlines. One common problem in the airline industry is that while passenger expectations are measured and molded by the marketing and sales departments, performance is generally in the hands of the operations department. Further compounding the problem, marketing and operations departments each utilize different criteria on which to base their decisions. Positioning CRM within one or the other of these two departments within an airline could result in limited use of the capability. An operations department's primary goal might be to minimize costs, while the marketing department's goal might be to attract and retain passengers. CRM is capable of assisting on both fronts. Ultimately, CRM is uniquely capable of measuring passenger satisfaction, which is a function of marketing expectations divided by performance. To maximize the capability of CRM, then, at the very least, marketing and operations departments would need to collaborate on a continual basis. However, to survive and thrive in an increasingly competitive environment such as the airline industry, effective collaboration and coordination of processes is required across many functions and business processes.[5] Consequently, it is worth considering a separate CRM unit at the corporate level, reporting directly to the CEO.

A better way to perceive CRM is to think of it as a process that manages customer expectations. In this context, CRM can be viewed as a way to maximize revenue opportunities, minimize customer costs, and not only improve the delivery process but also differentiate the service provided. Two key areas of managing revenue are targeting profitable customers and developing a clear, differentiated value proposition. Airlines are just beginning to make progress in this area. On the cost side, the current focus is on reducing sales and distribution costs whereas CRM initiatives can not only help to reduce costs in many other areas but they can, at the same time, improve the service provided to customers at different touch points.

Consequently, maximizing potential in terms of revenue, costs, and services should be a goal for the entire airline, not just one department. Full exploitation of CRM therefore requires airlines to be prepared to change their business strategies, processes, procedures and culture. Consider the processes at airports. They need to be redesigned to make the information collection, structure, and retrieval system more efficient. Airlines are just beginning to provide customer service agents with Windows-based pop-up systems. The change necessary is the need to shift the focus from operations to customer service. This is not to imply that airlines up to now have ignored customer service. Rather, airlines have not provided their employees adequate resources (information, skills, transactional capabilities, and incentives) with which to provide true customer service—in real time and at all touch points. Employees need to be enabled to in fact drive the CRM process at various touch points.

The second critical success factor relates to the means and quality with which information is collected and stored. Information must be correct, used correctly, customer focused, linked across all relevant databases, and mineable. This means that each data point must be tagged or indexed according to how it will be used. For example, when a customer books a reservation, a raft of data becomes available with that transaction. The particulars of the transaction must be segregated and identified appropriately (customer name, price, method of purchase, carrier, number of stops, departure and arrival time, and so on). Each data point must then be accessible so that when the system attempts to calculate how many passengers fly red-eye flights, it can call up all those reservations that were made within a certain time range. When devising a CRM process, then, an airline should identify first how the information will be used so that it can then flag each data point it will need in order to calculate customer behavior and needs. The customer-focused attribute of information relates to various aspects of customer care and the need to monitor useful customer activities on a continuous basis. Here is one example of customer-focused information relating to the use of airport lounges. Some airlines do not even collect information on the number of passengers using the lounge, let alone the names of passengers. And even the few who do collect the information on who used the lounge do not have any information on the facilities and amenities used by a particular passenger in a particular lounge.

The third critical success factor in CRM is the integration of the data from both external and internal sources. The need to integrate information from different sources cannot be overemphasized. Consider just the different channels of distribution available to the passenger: airlines themselves, traditional travel agents, websites of individual airlines, online travel agencies, and online discount channels. The proliferation of the channels of distribution combined with the disparate and disintegrated data sources within the airline makes tracking just the customer's purchasing behavior difficult, to say the least. Currently, in most airlines' internal sources, the data is not even collected at each of the passenger's touch points, let alone is it integrated and converted into meaningful information. Without the data, how is an airline to know if a customer is using an online discount vendor because of convenience or because of price? Consider also the process of tracking, claiming, and redeeming frequent flyer rewards. The organization responsible for processing the rewards does not know the value of each customer other than the status of each customer with respect to frequent flyer program value. How, then, can the success of a frequent flyer program be determined? Data relating to activity with sources external to the airline must also be collected, and it must be integrated with the entire CRM process so that it, too, is mineable.

The fourth critical success factor for CRM is that the airline must deliver its service on the passengers' definition of value. It is important to know how satisfied passengers are with each of the airline's value propositions (e.g., waiting time for reservations, baggage delivery, quality/quantity of in-flight food, and on-time performance). An airline must also know the level of importance a passenger places on each value proposition, for example, whether it is more important that the plane arrives on time than the ease with which the passenger booked the flight. Needless to say, the importance of each value proposition will vary by passenger segment. Leisure passengers are likely to value low-price tickets higher than business travelers.

Once an airline has this information, it can then compare each value proposition in each segment to reveal which proposition is most important for each segment. This comparison can be represented visually in bars. One bar can represent the satisfaction for each value proposition, and a second bar next to it can show the importance of each value proposition. With such a bar graph, the difference between the importance of value and the degree of satisfaction with that attribute can become readily apparent, as does the direction of the gap. The same exercise can be performed for satisfaction regarding each value proposition for the airline and the industry. Determining passenger satisfaction with each value proposition will enable airlines to focus their energies on providing service in the area customers in each segment value most.

Tracking such detailed value-based segmentation criteria and the associated costs require detailed passenger-related accounting systems similar to activity-based accounting systems. Then, in order to extract the necessary knowledge from such a database, airlines must employ such practices as data warehousing and data mining, and, ultimately, customer relationship management. Employing these more advanced data gathering and marketing techniques can enable airlines to offer services that are more targeted to the needs, behavior, and values of their customers.

The costs associated with developing such a data infrastructure may seem high, but the benefits would be well worth the investment. Consider, for example, tying seats on individual flights to special events such as sports and entertainment. This kind of offer is the beginning of one-to-one marketing, that is, creating customer-specific marketing offers. The technology is currently available that could allow such customer-specific offers to be made in real time. For example, if a customer has searched online for the lowest fare and is now looking at hotel options, the airline could make an offer on a low-priced room from a hotel partner. Such marketing initiatives could stimulate more traffic and yield greater profits. Providing such value-added service first of all reduces the commoditization that has long been associated with the airline industry. What is perhaps more immediately apparent is the effect such value-added services could have on pricing. If bundled services were available—plane tickets coupled with event tickets or with hotel reservations—those services would save time for passengers and provide one-stop travel shopping, and the price of the flight could then be set higher rather than lower, a strategy that could ultimately translate into higher profits. Offering such value-added or customer-specific services and achieving those attendant profits is really only possible through the use of customer relationship management.

The fifth critical success factor for CRM is that it should be applied across all channels and all functions. True quality service, for an airline, means offering products that meet the passengers' needs and provide attentive service as those products are delivered. In order for an airline to be able to offer this quality service, it needs to be able to differentiate its customers in terms of those potential products and services. Since CRM's purpose is to determine that differentiation, it can thus provide optimal value to customers both in terms of product, price, promotion, and distribution, as well as in terms of the interactions between the airline and the passenger at various touch points. Utilizing CRM to determine one marketing strategy or one component such as pricing therefore fails to take advantage of the range of capabilities CRM affords. Nor does it pass along those advantages to the customer.

The sixth critical success factor of CRM is to use it to develop service relationships with customers. Gutek and Welsh describe three types of service interactions, which they term relationships, encounters, and pseudo-relationships. In a relationship, the customer personally knows the service provider (a doctor, for example) and expects to interact with the same person again and again. In an encounter, the customer may know the organization (for example, McDonald's) but receives the service from whoever is available (the cashier at one franchisee's restaurant). In the pseudo-relationship, service is delivered in an encounter structure (chain drug store) but the delivery is made to feel like a relationship (a pharmacist who has access to information about other medications a customer may be taking and can therefore warn the customer about counter-indications).[6]

An airline simply cannot provide true one-to-one customer service, as in Gutek and Welsh's definition of a relationship. Given the multitude of passengers from around the world who are using the airlines all day every day all year long, in addition to complications resulting from flight delays or inclement weather, and the current lack of adequate systems, it would be impossible to devote one customer service representative to each passenger. However, airlines should be able to forge pseudo-relationships with their most profitable customers. That is, when a passenger contacts an airline to book a flight, the service professional assisting the customer should have access to that customer's profile that would indicate the customer's previous flights, his or her level of satisfaction with the services related to those flights, and the customer's overall profitability. The service professional could see immediately that the customer currently on the phone is, for example, a frequent business traveler who prefers higher-priced window seats and that the last two times he flew, his bags were lost. The service professional could then ask the customer if he would again like a window seat in first class and could also reassure him that the airline will take extra precautionary measures to ensure that his bags will not be lost a third time. In this way, the data will help the staff offer service as if he or she knew this particular customer, and the customer will in turn feel as if he has a relationship with this particular staff member. Ultimately, that pseudo-relationship with the service member will translate into the feeling of a relationship with the airline.

The final critical success factor is to recognize that the successful implementation of CRM is an evolutionary process with multiple phases where each phase delivers incremental benefits and builds a foundation for the next phase. This view is illustrated in Figure 3.5, adapted from a presentation made by William Brunger of Continental Airlines.[7] An airline must first start with a strong infrastructure as the foundation for the CRM initiative. The foundation infrastructure encompasses multiple customer data elements collected from different functions within an airline. Examples of data elements include passenger history with respect to both buying behavior and product delivery. Since it is almost impossible to have thought of all the data elements at this point, the foundation must remain flexible and scalable.

In the evolutionary process, benefits can actually start occurring as soon as data begins to flow in the system. An airline should not wait until the entire data warehouse has been completed and data mining capabilities have been evaluated and implemented. One airline, for example, began to notice almost immediately from the very initial data the degree to which fare inventory was being abused, both externally and internally. For example, suppose an agent books a group of 40 passengers as eight groups of five. This action could have a serious detrimental impact on revenue management control and overbooking policies. An airline will most likely make a different decision regarding overbooking involving a group of 40 vs. five groups of eight passengers. Similarly, an airline could make a different decision regarding overbooking based on which reservations were made before 11 September and which were made after 11 September. The show-rate is likely to be higher for reservations made after 11 September. Consequently, experience shows that benefits can start to accrue very early in the evolving CRM process. In the case of one airline, the very initial benefits actually paid for the entire cost of the data warehousing and mining system. Staff at this airline did not even know the potential uses of data until the system starting collecting the data and making it available to different groups within the airline.

During the next phase shown in Figure 3.5 (Customer Knowledge Management), raw data, contained in the customer information files developed in the infrastructure phase, is converted into actionable information. The customer files become more robust in this phase due to the combination of raw data provided by the customer in the first phase (such as seating preference) and data observed by the airline (such as the amount of time a passenger allocates between check-in and boarding).

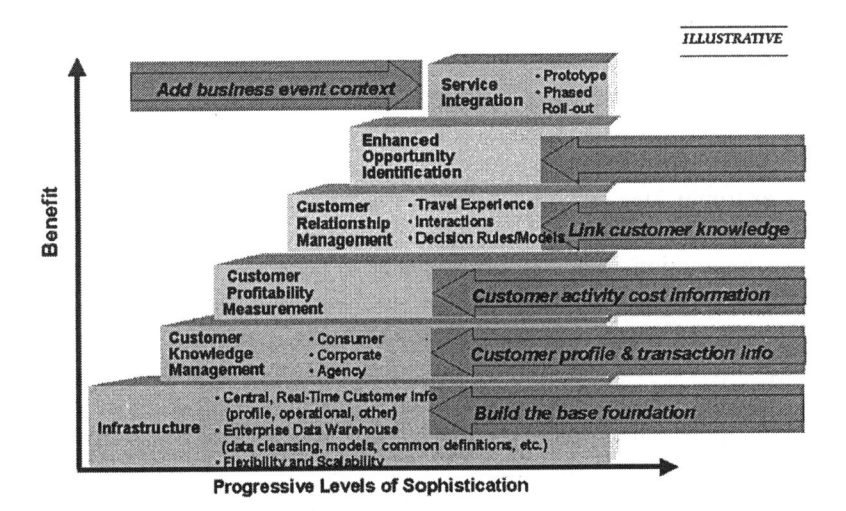

Figure 3.5 CRM—an evolutionary process
Source: Continental Airlines

The third phase shown in Figure 3.5 (Customer Profitability Measurement) contains data on the customer's value and the cost to serve the customer based on such information as the frequency and class of travel and the preferred channel of distribution. This phase essentially provides information of the potential revenue to be generated from this customer's business and the costs to be incurred in meeting the needs and wants of the customer. Such information is valuable in determining the resources an airline should commit to gain and retain a customer in the event he or she should defect.

The next phase shown in Figure 3.5 (Customer Relationship Management) allows the airline to track and monitor high-value individual customers or groups of customers. It is possible, for example, to identify an individual who usually travels once a month and has not shown up for three months or groups of passengers who fit a certain profile. The next phase (Enhanced Opportunity Identification) provides opportunities for an airline to generate more revenue from a customer based on the information available on the customer. Based on the existing situation, an airline, for example, may try to sell a higher fare seat on a nonstop flight or a membership in the airport clubs.

The final phase (Service Integration) enables an airline to serve a customer the way the customer wants to be served based on the stated and observed requirements of the customer. Some customers want to be recognized in a personal and friendly way—Happy Birthday, Mr. Jones, for example. Others may also wish to be recognized but at a more professional level—for example, "welcome back, nice to see you again."

This is, in fact, the true goal of CRM: to personalize the passenger's interaction with the airline. Though an airline cannot apply CRM on a true one-to-one basis, it can use the data that CRM makes available to gear products and services to the needs, behavior, and values of the airline's most profitable customers. Making such customer-specific services available is the future of competitive differentiation among airlines. Implemented correctly, with properly warehoused data that is made available in real time to customers, service professionals, and management alike, CRM will enable an airline to reduce costs and increase revenue. However, to achieve these objectives an airline must look at CRM not as a quick fix, but rather as a new way of doing business. And it is emerging technology that is the key to the cost-effective implementation of CRM initiatives. Emerging technology can not only link existing legacy systems but also provide the correct information and enable the appropriate service to be provided to each valuable customer at each touch point.

Notes

[1] Seybold, Patricia B., *The Customer Revolution* (New York: Crown Business, 2001), p. xv.
[2] Telephone interview with Dr. Lucio Pompeo, Associate Principal, McKinsey & Company, Zurich, Switzerland, August 2001.
[3] Interview with Michael Beckmann, Senior Consultant, Roland Berger Strategy Consultants, Munich, Germany, 11 December 2001.
[4] Baldanza, Ben, "The Evolution of Customer Segmentation," *Handbook of Airline Strategy* (New York: McGraw Hill, 2001), pp.361-72.
[5] Compton, Jason, "Mission Critical: Encouraging Collaboration," *CRM*, January 2002, p.37.
[6] Gutek, Barbara A. and Theresa Welsh, *The Brave New Service Strategy,* (New York: AMACOM, 2000).
[7] Brunger, William, "Customer Experience, CRM and Where We Might Go," Ohio State University Eighth International Aviation Symposium, Porto, Portugal, 30 May 2001.

Chapter 4

E-Business and its Application to Airlines

In the late 1990s, the failure of several high-profile dot-com firms using e-business strategies caused executives to question the value of e-business. It did become evident that the dismal financial performance of many of the first-generation dot-com firms may have been due to the result of an unrealistic inflation of their market capitalization (based on unrealistic projections of the value of customers) rather than the failure of e-business itself. Executives hesitate to write off e-business completely because it is in general a very customer-oriented practice, which, in this age of the customer, makes it attractive. Still, some executives remain wary of the potential risks of total immersion in e-business at this early stage.

Within the airline industry, the degree of this risk as well as the value of e-business varies from airline to airline depending on such factors as the stage of development of the airline, its business strategy, and the top management's view of the role of technology. As a result, airlines have a varying degree of interest in e-business. Toward one end of the spectrum, a few airlines are quite satisfied with a very basic Web presence. Toward the other end of spectrum a few airlines want to be at the leading edge of using emerging technology to find digital solutions to the triangular problem of customer expectations, operating costs, and competitive differentiation. At this time, virtually no airline wants to be at the bleeding edge of technology to totally transform itself into a full-blown e-business. This view is particularly important in the aftermath of 11 September that has caused airlines to reduce significantly staff and services in the short-term and look for a clear perspective in the long-term in light of not just the events of 11 September but also the industry's structural changes, for example, the consolidation process and the enormous success of low-fare airlines.

Despite variations among airlines' views of e-business, the airline industry as a whole possesses a number of characteristics that make it a natural candidate to benefit significantly from moving up the e-business leverage curve. These characteristics which e-business can improve are essentially the three points of the triangular relationship among operations, customers, and competition. See Figure 4.1. Each point of the airline industry triangle is currently being challenged, to varying degrees, by the following five forces:

1. *Deregulation* Varying degrees of deregulation of the airline industry in different parts of world have produced an industry that is highly competitive on major routes and major regions of the world. See he discussion in Chapter 1 on the growth of low-fare airlines in Great Britain.
2. *Globalization* Proliferation of the globalization has (a) made the industry even more competitive by bringing new competitors in the marketplace, (b) led to a tremendous increase in the breadth and depth of strategic alliances, (c) brought new suppliers, employees, and customers in the marketplace, and (d) led to a certain amount of consolidation within the airline industry.
3. *Customer expectations* Technologies such as the Internet and mobile communications have enabled other industries to reduce their operating costs and to fulfill the changing expectations of their high-margin customers. Customers now generally expect to receive a product or service when and how they want it. Customers are bringing those expectations to the airline industry, including more control over their travel experience and a greater focus on safety and security.
4. *Differentiation challenges* Commercial air travel is fundamentally a commodity business, which makes competitive differentiation difficult. Generally speaking, airlines use similar aircraft, offer similar service, charge similar prices, and use similar channels of distribution. Given these basic similarities in product, airlines have to find other ways to develop a sustainable competitive advantage.
5. *Operational and service complexities* Airline operations are inordinately complicated by internal challenges such as unionized crews and government-mandated maintenance requirements and by factors outside of the airline's control, such as weather, capacity of

the air traffic control system, and acts of terrorism. Airline customer service interactions are also complicated by the existence of so many customer touch points—points of interactions and transactions—in the total end-to-end service sought by the customer. And the degree of complexity is increasing in both service and operations.

These five forces are all pushing airlines to increase their operational efficiency, improve customer service, and boost competitive advantage. E-business provides some solutions to some of these problems, at each point on the triangle shown in Figure 4.1.

Before describing how e-business can benefit airlines, it is important to note that e-business need not be deployed on a total airline basis to be effective. Even at partial deployment, the industry has already proven its ability to sell online. Furthermore, airlines have already realized some of the basic benefits of e-business. Airlines have first of all enjoyed improved profits, either through increased revenues or through decreased costs, or a combination of the two. E-business has also improved service to passengers by making it more convenient to book reservations or to verify the status of a flight. Airlines can do more, however, to maximize the potential benefits e-business has to offer.

In order to reap these potential benefits, it is important first to understand what e-business is, both in general terms and how it can affect an airline's customers, the total airline in general, its partners, and its employees. Second, to implement e-business successfully, some adaptation to e-business is required of an airline. After defining e-business in terms of its effects, this chapter explores each of those essential adaptations.

Overview of E-Business

Description of E-Business and its Evolutionary Stages of Development

Some people think of e-business as another name for e-commerce. However, e-business is broader than e-commerce, which is generally related to the buying and selling of products and services over the Internet. E-business, on the other hand, encompasses the strategic use of information and communication technologies to interact with customers, employees, suppliers, and partners through multiple communication and distribution channels to maximize value for all parties.[1] See Figure 4.2. Let us examine some of the elements of e-business included in this description and portrayed in Figure 4.2.

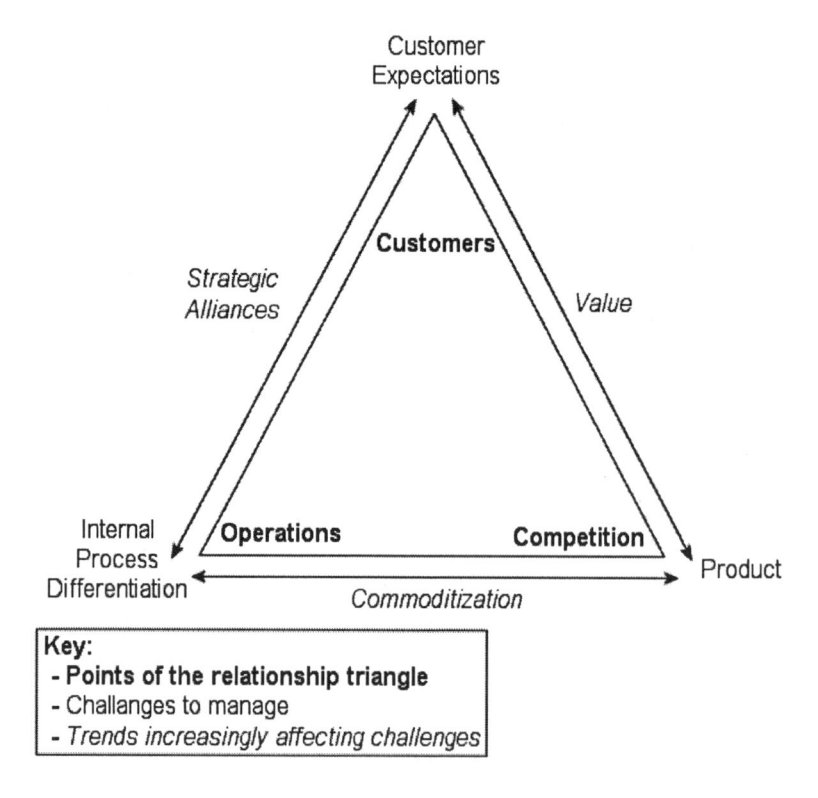

Figure 4.1 Airline industry triangle: relationships, challenges, and trends

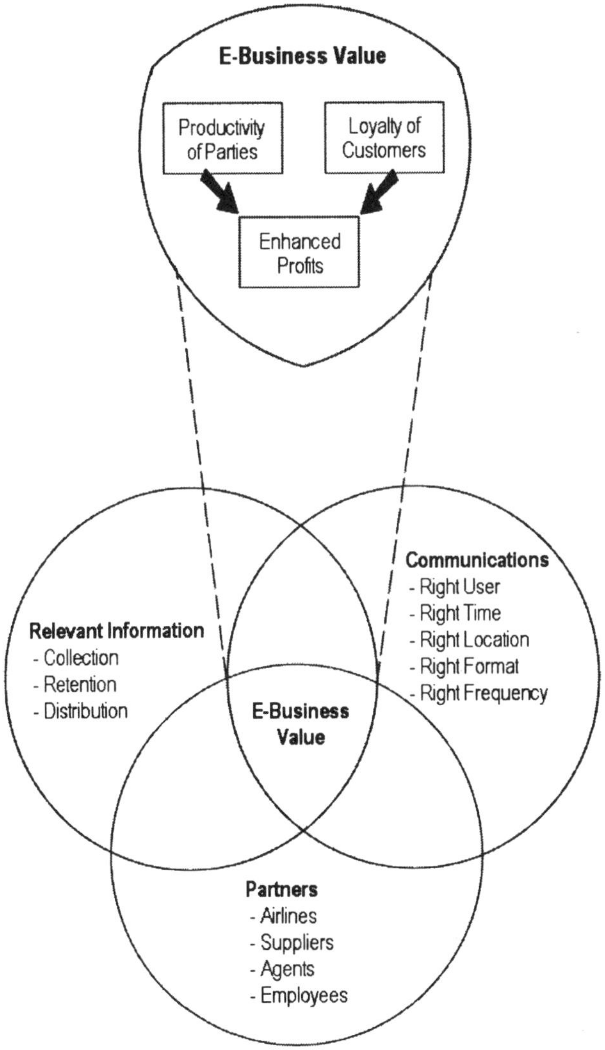

E-Business and its Application to Airlines 89

Figure 4.2 Value of e-business

1. *Use of information* Airlines collect vast quantities of data from both internal and external sources. This part of e-business requires not only the collection and retention of all relevant data at a central location, but also (a) the conversion of this data into timely and

actionable information, and (b) the distribution of the relevant information to all parties.

2. *Use of communications technologies* Information is of little value if it is not available at the right time, for the right user, at the right location, and in the right format. The kind of communications technologies employed can have a significant impact on the information it is intended to manage.

3. *Partners* For airlines, partners include traditional entities such as employees, suppliers and other airlines in strategic alliances, as well as other parties, such as travel agents, airports, and airport immigration and customs agencies.

4. *Effective communications through multiple channels* A passenger, for example, may receive information from an airline call center, a strategic alliance partner, a travel agent, an airline's website, or any number of other airline online services. Each of the parties who may be providing information to the passenger should also have access to each other's information so that the data each party keeps can be verified and updated. The sharing of this information also needs to occur on a timely basis so that everyone has consistently correct data at the same time. The end objective is to utilize information that is correct, timely, consistent and universally available to all parties.

5. *Value* The value of e-business is increased as the productivity of all parties increases and as the loyalty of customers increases. Increased productivity and increase customer loyalty can, in turn, translate into increased profitability.

Because this description of e-business is broad in its scope, e-business can be implemented at varying stages of business processes. For example, one group of e-business strategists has offered a classification of those stages in four types: channel enhancement, value-chain integration, industry transformation, and convergence.[2] Let us examine these four stages in the context of an airline.

1. *Channel enhancements* During this stage, an airline would deploy e-business technology basically for sharing information and conducting e-commerce, which primarily means buying and selling products and services over the Internet. In this context, the reader should not confuse channel enhancement with the airline industry's

concept of distribution channels. The broad description could also include an airline's purchase of the products and services it needs to produce its own services.

2. *Value-chain integration* At this stage, an airline can leverage its e-business initiatives—for example, collaborative planning—to improve the relationships among an airline and its value-chain partners, such as vendors and suppliers. These relationships can be improved in terms of increased efficiency and reduced costs, so that value is increased for all parties.

3. *Industry transformation* This stage integrates the core activities of different companies to maximize shareholder value for all companies. For airlines, the parties involved could presumably be all strategic alliance partners, or an airline and its vendors/suppliers, or all airlines and all their vendors/suppliers.

4. *Pan-industry convergence* Here, the group envisions the integration of activities of companies across different industries. Again, within the airline industry, one could envision the integration of all travel-related industries, such as airlines, car rental firms, hotels, restaurants, ground transportation firms, entertainment organizations, and credit card companies.

Given the varying levels of implementation that are possible, it is up to each airline to decide on the stage to which it wants to climb the e-business ladder to find solutions to the triangular issues mentioned above—customer expectations, operating costs, and competitive differentiation. Consider, for instance, the following three options that address the potential use of e-business relating to one or more of the three corners of the triangular industry dilemma (Figure 4.1) than the four stages of implementation discussed above:

One airline may decide to continue to provide basic transportation in the traditional manner and use only those components of e-business technology that reduce its operating costs by increasing its internal efficiencies. One way an airline could implement this level of e-business strategy would be to use sophisticated fare-auditing and revenue accounting systems to identify actual revenue generated from tickets sold by the airline as well as by partner airlines and travel agents. The use of these two systems addresses the component of revenue enhancement/operating costs. Regarding the other two components—

customer expectations and competitive differentiation—this airline may decide to lower fares and to stress courteous service instead of investing in either high-touch or high-tech service infrastructures.

A second airline may decide to focus on the customer service corner of the triangle and to offer high-tech rather than high-touch solutions, such as by checking in passengers using the traditional process and by offering self-service check-in machines, or remotely with the use of hand-held mobile devices. Such an airline would, presumably, make its business case for the deployment of e-business technology along the following lines. The additional costs of technology would be more than offset by the lower labor costs, resulting in an overall reduction in operating costs. Moreover, this airline may view that customer service would be improved by providing, for example, a choice to check-in options. Even though the competitive differentiation may not be sustainable, it would exist at the beginning and if the airline continued to deploy the latest technology, eventually the airline could become known for continuous innovation. A third airline may decide to climb really high on the e-business ladder by focusing on the total customer experience through variables that are, to varying degrees, under the control of the airline—operations and customers.

E-Business and its Effects on Four Constituents

To whatever degree an airline implements e-business, it will have a different, yet specific effect on customers, the airline in general, partners, and employees. We will explore the effects on each of these entities in turn.

Customers E-business activities can be applied to all points of the total customer experience, from pre-pre-flight to post-post-flight. In the pre-pre-flight phase, the customer obtains information, selects an airline, and makes reservations. The pre-flight phase involves all phases of check-in and boarding. The in-flight phase relates to the customer's use of cabin staff, facilities, and services. In the post-flight phase, the customer deplanes and claims baggage. In the post-post-flight phase, the customer may need to get in touch with the airline regarding such issues as those relating to baggage and mileage credits in frequent flyer programs.

E-business can be applied at any one of these points in the travel experience with three general goals in mind. The first is to make the customer's access to various components of airline service as convenient and safe as possible. The second goal is to give the passenger as much control as possible over any component of the air travel process. The third goal is to personalize service for the passenger as much as possible. All three goals are consistent with the discussion on CRM—a component of e-business—presented in the previous chapter.

Consider the goal of convenience. Suppose, for example, that on a Sunday evening, in a small town in the USA, a passenger wants to know the schedule of a small airline based in Europe or in the Asia-Pacific region and make reservations. There are no travel agents with offices open at this time and the airline the passenger wants to fly with may not even have a call center in the USA, let alone an open call center staffed with live customer service agents. Website technology—a component of E-business—can make it possible for the passenger to obtain flight information and make reservations at a time that may not be convenient for the airline, but is convenient for the passenger.

Consider now the goal of control. Assume that a flight is scheduled to arrive late, thereby causing the passengers to miss their connections. The standard procedure is that after the late flight arrives at the gate, some passengers making connections rush off, hoping to catch their next flight, some try to work with the limited agents available on the ground to find alternate flights, some passengers call their agents or offices, and some try to contact the airline's call center to get re-booked on other flights. Service recovery—another component of e-business technology—would provide to all passengers while still on board the late flight information on the status of connecting flights as well as all available options for alternate connections. Providing this kind of real-time information to passengers that allows them to choose among options gives the customer greater control over how the problem is resolved. The passenger's stress is reduced, and valuable airline staff resources are considerably less taxed.

Finally, let us consider the personalization goal. E-business allows an airline to act on the individual preferences of its customers. If a passenger receives e-mails offering very low-priced distressed seats, those offers are not meaningful if the passenger does not have any interest in that city or does not have the time available to take advantage of the offer. The passenger may in fact regard those e-mails as a waste of time and an annoyance, rather than a useful feature. On some airlines' websites, however, passengers can now record their preferences and notification instructions so that the site displays only offers that match the passenger's criteria. E-mails generated by a match between customer preferences and available flights, could also be sent to notify the customer of flights that meet those specific travel needs. The customer would then spend less time searching for flights and fares, and the airline would be more likely to capture that customer's future business.

Airline websites themselves can also adopt greater personalization capabilities. Currently, airline websites in general work as a system works, not the way passengers think. For example, they provide information on a round-trip basis, rather than providing more flexible information about flights to one city and returning from a different city. Some websites rank the customer's search results with respect to airline-specific criteria such as direct flights or the least amount of layover time at intermediate cities. The passenger, however, may prefer to rank flights by other, passenger-specific criteria, such as the number of frequent flyer miles airlines offer for that trip. Airlines are beginning to expand the features of their websites to allow for greater personalization, but even more could be done to make the sites passenger-oriented rather than airline-oriented. The more personalized the information, the easier it is for a customer to find the flight that best suits his or her needs for fares, scheduling, comfort and flight-related benefits. The customer who is taking a flight that most closely meets all of these needs will be the most satisfied.

Airlines The Internet, along with other such technologies as mobile communications, is redefining the total travel process and total passenger travel experience. During each phase, from pre-pre-flight to post-post-flight, an airline can implement procedures and systems to learn more about the customer, make meaningful marketing offers, influence and educate the customer regarding product features, and manage its relationships with an individual or a group of customers in a specific segment. The Internet has opened new ways for airlines to enhance their revenue, reduce their operating costs, and improve the service provided to their customers.

E-business technology has enabled airlines to increase revenue by attracting new customers through their websites, by filling up more of their capacity through Internet-based pricing schemes that harmonize capacity and demand, and by making customized marketing offers to targeted customers. CRM initiatives, described in the previous chapter, also offer significant potential for increasing revenue, since these initiatives maximize the retention of existing customers.

Airlines' use of e-commerce for electronic procurement and supply-chain management has also reduced costs significantly. The Internet is further reducing costs since it provides a new channel of distribution that is available 24 hours a day and does not necessarily require travel agents or third-party computer reservation systems.

Airlines can do more to reduce costs using e-business technology specifically in the operations department. Chapter 2 contained a discussion on various aspects of operations across different functions. Consider, for a moment, just the case of maintenance and engineering. In some areas of maintenance, such as line maintenance, the cost of labor is much higher than the cost of materials. In other areas, such as engine overall, the cost of material is much higher than the cost of labor. In both cases, the deployment of e-business technology can reduce maintenance costs. Labor costs can be reduced by making the maintenance staff more productive by making, for instance, technical information more readily available. In the case of materials, costs can be reduced by lowering purchasing (product price and cost of transactions) as well as by reducing the cost of holding inventory of spare parts.

Similarly, costs can be reduced in other functions ranging from catering to crew uniforms. Consider, for example, an airline with US$8 billion in total operating costs. Even if one assumes that 75 percent of the total operating costs fall in categories that cannot benefit from e-commerce (for example, labor and fuel costs), that would still leave US$2 billion where substantial savings could be achieved through e-commerce initiatives. A 10 percent savings in the remaining operating costs would lead to an annual savings of US$200 million. The upside potential for savings is substantial once an airline starts examining the areas that make up the 75 percent and resolving some of the problems within that area using business-to-business transactions. Examples include the difficulties surrounding the establishment of lines of credit, establishing credibility about the legitimacy of the other party, dealing with cumbersome customs requirements, and problems relating to currency for payments.

Achieving such savings requires companies to change their business processes at one or both ends of the supply chain. Examples of changes include the installation of intranets and extranets and other communication systems among all parties to enable collaborate planning through the Web and outsourcing of non-core activities. One intriguing example suggested is to use the Internet to improve its internal processes to reduce internal costs and improve service provided to customers at the same time.[3] The suggestion is for an airline to set up a Web page on its server for each aircraft. This Web page can contain both the historical and the most current information on each aircraft. Although much of this kind of information is generally available in an aircraft's logbook, the information on the Web page could be mined quickly to identify patterns that may involve particular crews, maintenance technicians, and parts suppliers. Such a change in process can enable an airline to convert raw data into timely and actionable information that, in turn, can enable an airline to reduce maintenance costs and at the same time improve customer service by improving on-time performance.

Partners The shifting dynamics of the airline industry have been changing the types and importance of partnerships. For example, the trend in North America is to rely less on travel agents. The trend to rely more on airline partners within strategic alliances is, however, worldwide. The change in partnerships and their importance has led to an even greater need to integrate the activities among the partners within an airline strategic alliance to provide a seamless travel experience for the customer and at the same time to reduce operating costs for all members of a partnership. Although e-business technology can help airlines within a partnership achieve both of these objectives, the task is not easy for the following reasons:

1. The changing dynamics of the airline industry are leading individual airlines to move from one alliance to another. Consequently, each airline within a given strategic alliance needs to be reasonably sure of its long-term participation in that alliance. The costs of changing technical systems to comply with different alliance partnerships are enormous.
2. While the basic objective of an alliance is to collaborate, in the airline industry there is still significant competition among the partners within alliances. Consequently, airlines are reluctant to openly share their proprietary information.

3. Airlines work with old technology systems that are accurate and efficient to deal with old ways of handling passengers. In today's environment, the legacy systems are not capable of accommodating the needs of airlines or their customers in a technology-rich environment encompassing online reservations, e-ticketing, mobile communications, and CRM processes. The main system of each airline handles literally hundreds of applications. Even if one assumes that each airline is willing to share these applications within the alliance partnership, up until now it has only been possible to integrate technology at the bilateral level between two airlines at a time. Technology is now just beginning to enable airlines to integrate activities at the multilateral level.[4]
4. Even when technical problems seem to be resolved, differences among airlines in culture, products, processes, strategies, skills, and trust need to be bridged, as does the level of commitment to the membership within a given alliance, and differences in the vision and perceived value of e-business initiatives.

Despite these difficulties, each partnership will succeed in the adoption of some level of e-business initiatives because: (a) it is a generally accepted view that strategic alliances among airlines are the only way for major traditional airlines to survive in this increasingly globalized world; and (b) emerging technology may itself be a way for airlines to compete and collaborate at the same.

Employees It should not be a surprise to read that employees are perhaps the most important component of the strategy to resolve the three critical issues mentioned earlier—customer expectations, operating costs, and competitive differentiation. Similarly, it should not be any surprise that the critical component of e-business is once again employees. E-business technology can enable and reward employees by (a) including them in the information loop so that they have some understanding of the airline's overall strategies and policies, and (b) by providing the employees with tools and systems to enable them to at least meet, if not exceed, customer expectations.

Let us consider the importance of employees from just one perspective. Figure 4.3 shows the results of a survey conducted by J.D. Power and Associates and the *Frequent Flyer* magazine. This chart shows the breakdown of the passenger satisfaction influenced by different factors. This information comes from a survey of frequent flyers traveling on major U.S. airlines. Virtually all of the satisfaction is associated with nine factors. Although to varying degrees, all nine factors have a labor component associated with them. Take, for instance, on-time performance, the single factor responsible for passenger satisfaction. There is a broad spectrum of employees whose activities affect on-time performance of an airline, for example, maintenance, crew, gate agents, aircraft cleaning staff, catering staff, ramp agents, dispatch staff, and baggage handlers.

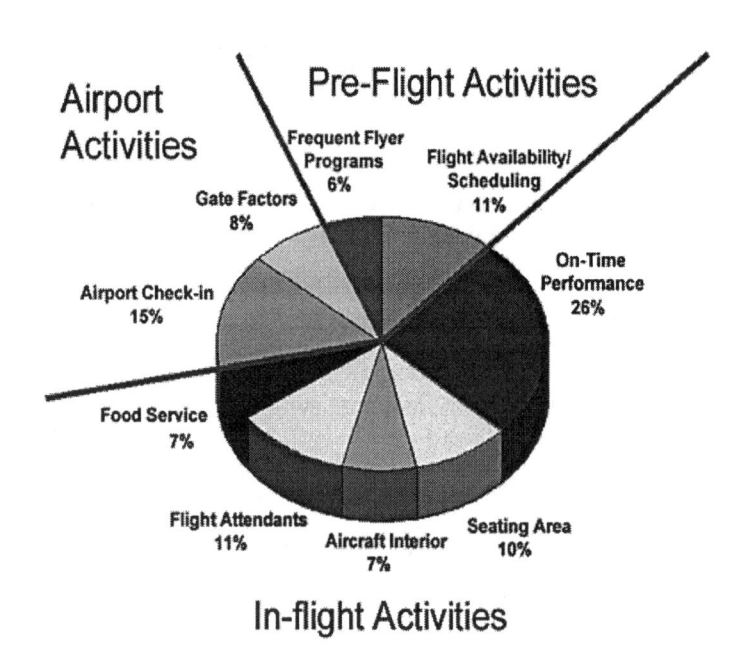

Airport Activities
- Gate Factors 8%
- Airport Check-in 15%
- Food Service 7%

Pre-Flight Activities
- Frequent Flyer Programs 6%
- Flight Availability/Scheduling 11%
- On-Time Performance 26%

In-flight Activities
- Flight Attendants 11%
- Aircraft Interior 7%
- Seating Area 10%

Figure 4.3 Nine factors drive overall airline satisfaction
Source: J.D Power and Associates and *Frequent Flyer* Magazine, 2000

E-business initiatives can help coordinate the activities of all. The employee-focused e-business initiatives provide employees access to people, information (internal and external), and services to not only communicate but also to learn and to collaborate—the core aspects of knowledge management, discussed in Chapter 7. The customer-facing employees can also play an important role in educating the customer about some aspects of the product and service, for example, why sometimes an airline cannot provide information on the status of delayed flights. Knowledge management, therefore, applies for both passengers and employees. As for employees, every airline has some employees who are exceptional, may they be in maintenance, scheduling, systems, fare auditing, or the complaints department. It appears that the knowledge they possess goes with them when they leave work at the end of the day. The e-business capability provides a way to manage that knowledge for the overall good of the airline and its customers. It should not be necessary, for example, to discover a problem multiple times. If an agent at an airport counter has found a way to solve a particular problem, all agents should be made aware of the fix immediately.

It is important to keep in mind that to improve the total travel experience of the passenger and improve the productivity of the system, there is a need to coordinate the activities of three important components of the aviation system—the airline, the airport, and the air traffic control system. It is not inconsistent with the material presented in this chapter, namely, that e-business technology can enable the coordination among all groups—customers, employees and partners. Partners can now include not just other airlines within strategic alliances and other travel-industry partners in the complete travel chain (hotels, car rental companies, credit card companies, and so forth) but also airports and possibly even some aspects of the air traffic control system. It is the inclusion of all these components that would lead an airline to move to the fourth stage of e-business development mentioned above, namely, convergence.

Key Drivers of E-Business

E-business initiatives are not a way to reinvent an airline. Rather, e-business initiatives should be integrated into an airline's core business to redesign the core business to provide higher levels of customer and employee satisfaction. The implementation of e-business requires, not just the use of emerging technology, but also a clear vision statement, a redesign of business processes, the development of new performance metrics, and finally, a realistic transformational process.

Vision

Airlines have tended to be product and operation-centric and not customer-centric. Consequently, they tended to look for competitive advantage through product-based systems rather than through customer-based systems. Consequently, the first driver of e-business is to have a vision to transform the airline organization from product and operation-centric to customer-centric. The purpose of e-business can then be viewed as a way to develop competitive advantage by focusing on the creation of value for the customer with respect to his or her total travel experience that can be enhanced by the aforementioned three ways. First, a customer should be able to conduct business with an airline with great ease. Second, the airline should allow the customer to be in control of as many interactions and transactions as possible. Third, an airline should implement as many ways as possible to personalize the service.

Technology

Initially e-commerce entered the marketplace with a focus on business-to-consumer (B2C) transactions. This was followed by business-to-business (B2B) and then business-to-employee (B2E). Then, with the proliferation of wireless phones, attention became focused on mobile-commerce. Mobile-commerce promised interactions and transactions anytime and anywhere. Now, the trend is for customers to buy not only anything, anytime and anywhere, but also using any device. This phenomenon is becoming known as u-commerce, universal or ubiquitous commerce.[5] The u-commerce phenomenon integrates the financial aspects of the transaction, that is, the ability to pay for a product or service, not just anytime and anywhere, but with the use of any mobile device. However, all these phases of commerce are related to technology that provides real-time connectivity. In addition, there are two other key technologies needed to transform the airline business into an e-business—middleware technologies that integrate new applications (such as CRM) with an airline's traditional legacy systems (reservations, departure control, and so forth), and data warehousing and data mining technology.

Real-time connectivity technology has two major components—mobile technology and Web technology. Both of these components provide an enormous opportunity to manage proactively customers, employees, partners, and operations in order to improve customer service, reduce operating costs, and achieve a competitive differentiation. A broad spectrum of mobile devices already exists for internal and external operations. Let us take a few examples of these devices and their applications for e-business within the airline industry.

Personal Digital Assistants (PDAs) or mobile phones can be used to order tickets and receive a PNR (Passenger Name Record) as a small message. This eliminates the waiting time involved in calling an airline reservation center, particularly during busy time periods. Next, these devices can be used to send messages to an airline's reservation and departure control systems indicating that the passenger is at the airport—a message that can start the process of checking-in the passenger. The passenger can receive a bar code on his or her mobile unit that can then be used to process the passenger through the airport.

Available technology can enable an airline to send a given piece of information to a select group of passengers on their hard-wired or mobile devices. Suppose a flight is cancelled as passengers are checking-in for this flight. Suppose also that the agent checking in passengers notices that the next flight has only a few seats available. An airline can inform selected passengers (based, for example, on their status in the frequent flyer program) from the cancelled flight that they have been accommodated on the next flight. Such information can appear on the mobile devices of the affected passengers while they are still standing in the check-in line.

It is also possible for a hand-held device to be loaded with credit/debit card account information that can be sent to a terminal to pay for goods or services. On approval of the transaction, a digital receipt can be transmitted back to the hand-held device, which can later be printed, if desired.[6] This improves the service provided to the customer and lowers the costs for the seller. With this capability, it is possible for a passenger not only to make a reservation using such a device but also to pay for the ticket, receive a receipt and a PNR number, check-in and receive a seat assignment, a boarding card, and the gate number. The technical feasibility of such an application can easily be seen from the experience of the DoCoMo's iMode service in Japan that allows its 20 million subscribers to download music, shop and send instant messages.[7]

Blue Tooth is one type of technology that enables personal portable devices to interact with stationary devices and with each other using a short-range wireless communication system. Some airlines have already begun to use at airports roaming agents equipped with hand-held mobile devices that are connected to an airline's reservation system. These agents not only help passengers waiting in check-in lines but those anywhere at the airport such as security points, airport lounges, congested gates, and baggage claim areas. At the present time some of these devices have limited access to the central databases, for example, an airline's flights for that day of travel as well as limited information on the customer files. Nevertheless, even this limited information can help the agent provide personalized service at various touch points. Similar types of hand-held systems can be used by flights attendants in flight to provide passengers with information on connecting flights that may be delayed.

The other major component of real-time connectivity technology is Web technology that is needed to design websites. There are four basic considerations relating to the design of a basic website (information): content, functionality, user friendliness, and style. However, before taking into consideration each of these attributes, it is necessary to identify the intended user, for example, the general public or the frequent flyer. Content is important in that the user needs to be able to obtain up-to-date information on products and services as well as the airline. Functionality means that the user should be able to get the information in his or her desired format and be able to download the information in a variety of formats. Users should also be able to communicate with the airline if, for example, they want to ask questions or provide feedback on the website. The user friendliness attribute refers to the ease and the speed with which a user can find the information. Style refers to the use of graphics, color schemes, type and size of print, and adherence to local customs and practices.

After these basic four considerations comes more sophisticated requirements such as interactive applications that enable a user to check the status of things (frequent flyer miles, availability of a seat or an upgrade) and perform transactions (make reservations and buy tickets). A user may also want the ability to be connected to a call center to access a person. The next step might be to integrate the website with other channels of distribution and sources of information. The final step would be to personalize the website for key customers to make it easier for them to do business with the airline. However, in the personalization process, the user also expects the airline to protect the information available on the customer.

Web technology is also required to provide passengers with Web access at each seat and its connectivity to a station in the cabin-crew area. The availability of this connection can enable the cabin crew to provide personalized service to passengers (depending on the passenger's class of service, length of flight, and relationship with the airline), collect information about the passenger, and improve productivity of cabin staff. Web access at the seat can also be used to communicate relevant information to the passenger such as gate assignments of connecting flights or the status of the bags if the bags did not make the flight. In the later case, this information lets the passenger know not to go to the baggage claim area, but rather, to file the information needed by the airline to track the bags.

The major real-time connectivity technology limitation relating to in-flight entertainment and services at the present time has been the size and cost of bandwidth between the ground and the airplane. This limitation has restricted the use of in-flight entertainment such as live television, on-demand movies, and services such as access to e-mail and Internet surfing. However, emerging technology is expected to enhance the availability of these services that would (a) increase customer satisfaction, (b) have the potential to enhance revenue by charging fees for these services, and (c) improve productivity of in-flight staff.

One major problem with technology that is holding back the deployment of e-business is that a number of legacy systems (such as reservation, inventory, and departure control systems) are inflexible. They were initially designed to reduce operational costs through the use of operational optimizing techniques. Later, airlines' focus changed, for instance, to increasing market share with the use of, for example, strategic alliances. Now e-business focuses on customer retention through the use of, for example, CRM initiatives. Unfortunately, it is not easy to integrate the new applications with the legacy systems. Consequently, middleware technology is needed to enable legacy systems to execute new applications.

The final component of technology needed to implement e-business initiatives is the availability of a comprehensive data warehousing and data mining system. The need for this system was mentioned in the previous chapter within the context of CRM and will be discussed again in Chapter 7. At the risk of repetition, the rest of technology is fairly useless without information technology such as:

1. availability of information on a customer in virtually real time at every touch point
2. capability to mine the database to determine key patterns of customer behavior
3. capability to transmit the information to employees and devices at each touch point, again, virtually on real-time basis
4. capability of measuring the profitability of different categories of customers with respect to the revenue they provide and the services they require at each interaction point to obtain the satisfaction that they need.

Processes

The strategic use of information and communication technologies is necessary to provide service to the customer at the time and in the manner desired by the customer. However, this use requires (a) redesigning some existing processes, (b) integrating some other existing processes, and (c) developing some totally new processes. The following are some examples of all three types of processes.

An example of a process that needs to be redesigned is providing information on the status of a delayed flight. While passengers can understand the reasons for delay, such as mechanical failures and weather, they tend to get irritated when they are given, for example, information on a three-hour delay in 30-minute increments. Although it is understandable from the viewpoint of the airline why the process works that way (due to the uncertainty involved), it is still frustrating from the viewpoint of a passenger who is unable to plan. One way the process could be redesigned is to provide the information in a way that says the flight will not leave prior to this time but could be as late as this time. This reduces a passenger's anxiety and level of stress. Business-to-business activity is about reducing costs by optimizing the amount of products and services needed to operate an airline. It may also require a redesign of internal processes, reconfiguration of arrangements with suppliers, as well as strategic outsourcing.

Channel integration is an example of process integration. The lack of consistency due to the existence of dysfunctional channels is a major cause of passenger dissatisfaction. For example, a person getting the information and buying a ticket from a travel agent may get a very different interpretation of the product than one envisioned by the airline and its brand development department. Channel integration can improve customer service and reduce operating costs. Consider, for example, the productivity of a call center. It can be increased significantly by introducing not only Web technology but also e-mail and connectivity with the website. For example, a call center should be able to assist a customer by using its website.

Finally, an example of a new process, that some airlines have already instituted, is providing weather information on TV monitors at gates that shows weather patterns at the current airport and the destination airport. This information helps the passenger have a better understanding of any delays that might be announced and any changes in the passenger's plans affected by the weather.

Metrics

As mentioned above, many airlines continue to be product focused. Some who claim that they are customer-focused tend to be focused on their frequent flyers. As stated earlier, not all passengers are profitable and not all frequent flyers are profitable. Thus, there is a need to first focus on those passengers that are profitable and those who could become profitable by either increasing the revenue they provide or reducing the cost to serve them. As it was mentioned in Chapter 2, airlines are in serious need of new metrics to determine customer profitability. For example, it is well known that it costs more to acquire new customers than to retain existing customers. However, there is very little hard data on the costs relating to customer acquisition, retention, and growth for various customer segments.

Earlier mentioned were three sub-components of customer service that would improve the experience of customers: convenience, control, and personalization. However, while customer service is an important component of e-business initiatives, there is a need for new metrics to measure the costs and benefits of additional convenience, control, and personalization. The development of such metrics requires feedback from passengers at granular level on such factors as the importance they attach to various features, such as waiting in line to check-in, delays, and self-service capabilities.

E-business enthusiasts suggest, for example, that an airline can provide its high-value passengers dedicated e-mail addresses and personalized Web pages on its server. The e-mail can be used to communicate directly with these passengers. Web pages can be used by passengers to store their personal preferences for an airline to communicate targeted marketing initiatives as well as to monitor the usage patterns of various passengers. Some analysts even suggest the possibility of an airline assigning a customer service *bot* (an intelligent software agent or possibly even a human) to be connected to an individual high-value passenger's Web page on the airline's server. This agent could perform and monitor specific functions and communicate the results back to the passenger.[8] Again, there is a need for new metrics that measure the cost and benefits of such emerging customer service initiatives.

As mentioned earlier, e-business initiatives call for systems that (a) include employees in the information loop, and (b) enable employees to provide the service expected by different segments of the marketplace. Again, there is a need for metrics that measure the cost and benefits of such systems that measure employee satisfaction.

Next, there is a need for metrics to measure the performance of processes. For example, some e-business initiatives envision passengers to deal with systems rather than people. Staff are merely present to intervene in the event of a failure of the system or if the interaction takes place outside of the bounds established by management. For example, if it takes a passenger more than so many minutes to make a transaction, either a person can intervene or provide the passenger with a financial reward— certain number of dollars off the price of the next airline ticket. Once again, there are no metrics for evaluating the economic viability of such initiatives.

Finally, since the establishment of a successful e-business environment requires the existence of collaboration, it is important to set up an incentive system that rewards group effort.[9] New metrics are needed to measure the results of and rewards for collaborative planning, for example, in the case of strategic alliance partners in such areas as network design, customer care, and alliance performance measurements.

Transformation Process

Even though some airline managements do see value in adopting the e-business initiatives, they feel uncomfortable with its implications with respect to the need for initiating change and not so much the needed investments in technologies. This is natural and the fear of change can be overcome by adopting a modular approach. Let us assume that the airline is a traditional business and is just thinking of initiating e-business.

First, it is essential to appoint a project team to work on the e-business project. The team should be comprised of appropriate business professionals as well as technology professionals. The need for both types is critical. The technology professional, for example, may examine only the cost side or push technology for the sake of technology. The business professional may see an item not just as a cost center but as a potential revenue- or value-generation opportunity. A call center, for example, can evolve into a marketing center. In the past, call centers have been viewed in the context of cost centers rather than revenue-generating opportunities—a possibility by providing personalized service through proactively collecting actionable information from existing and potential customers. Finally, the team must consist of people who view the e-business initiative as "business as usual." As Tom Nunan of Cathay Pacific points out e-business initiatives require "new skills and fresh thinking—people who are not satisfied with status quo."[10]

This team can identify which e-business applications are important (B2B, B2C, or B2E) for its business strategy. In order for emerging technology to be adopted, the applications need to be relevant, use needs to be simple, and costs need to be commensurate with benefits. The mere existence of feasible technology does not guarantee it adoption by an airline or a passenger. It has to be relevant and cost-effective. Relevance means, for example, making a customer's life easier, simpler, and safer. It needs to present a clear business case and a value proposition for the passenger.

This airline can begin by adopting those components of the e-business that relate only to internal operations. The implementation of intranets can improve internal communications. In the past, for example, many pilots had very little contact with management. Some lived very far from the cities from which they flew their airplanes. They flew in or drove in, operated their flight, and went back to their homes or other businesses. The intranet provides an enormous opportunity for improving internal communications as well as staff productivity.

Next, the airline may want to launch a website, which can be more than just online presence or just another distribution channel—it has the potential to redefine the business model based on the information obtained on the airline's customers' profiles and their needs and values. However, as mentioned above, the development of the website should be undertaken in stages. At each stage, the airline can evaluate its rate on return on investment for each initiative before proceeding to the next stage.

There is no denying that adopting emerging technology has its risks; but it also has its rewards. Gaylon Howe of VISA International gives good advice with respect to the adoption of technology, "think big, start small, and scale fast." Thinking big means being visionary and recognizing potential. Starting small means testing the systems. Scaling fast means expanding the scope and scale to take advantage of the opportunity.[11]

Notes

[1] Siebel, Thomas, M., *Taking Care of eBusiness*, (New York: Doubleday, 2001), p.3.
[2] Deise, Martin E., Conrad Nowikow, Patrick King, and Amy Wright, *Executive's Guide to E-Business: From Tactics to Strategy*, (New York: John Wiley & Sons Inc., 2000), pp. xv-xxvii.
[3] Methner, Bruce E. and Christopher J. Rospenda, "Airline Strategy in a Digital Age: What Does "e" Mean to Me?" and Butler, Gail F. and Keller, Martin R. (eds), *Handbook of Airline Strategy*, (New York: McGraw-Hill Companies Inc., 2001), p.398.
[4] "A true test of friendship," *airline info tech: information management strategies*, Summer 2001, pp.16-18.
[5] Schapp, Stephen and Richard Cornelius, "U-Commerce: Leading the New World of Payments," VISA International and Accenture White Paper, July 21, 2001, p.1.
[6] Ibid., p.2.
[7] Ibid., p.3.
[8] Methner, Bruce E. and Christopher J. Rospenda, opcit, p.393.
[9] Sawhney, Mohan and Jeff Zabin, *Seven Steps to Nirvana: Strategic Insights into eBusiness Transformation*, (New York: McGraw-Hill, 2001), p.280.
[10] "Business as unusual," *airline info tech* (Spring 2001), p.20.
[11] Schapp, Stephen and Richard Cornelius, opcit, p.13.

Chapter 5

Opportunities Driven by Emerging Aircraft Technology

As the airline industry enters a new century, it must constantly adapt to new challenges. These include the world moving more and more toward a global economy, the proliferation of liberalization, the increasingly constrained aviation infrastructure, continued growth rate of air travel expected in the next 20 years, innovations resulting from emerging aircraft technology, and more recently the events of 11 September. This chapter discusses the role of aircraft technology helping the airline industry face these challenges and capitalize on opportunities. The first section deals with long-haul international markets and the role of the recently launched Airbus A380 and the Sonic Cruiser concept proposed by Boeing. The second section discusses the impact of regional jet aircraft in the short- and medium-haul markets. The final section discusses transportation by private aircraft—the phenomenon growth in the jet fractional ownership business as well as some novel ideas being examined for personal air transportation.

Long-Haul International Markets

The Airbus A380

When Boeing introduced its 747 in 1970, it resulted in an increased capacity of almost 150 percent from the then existing large aircraft such as the Boeing 707 and the Douglas DC-8. When the full-length double-deck fuselage A380 enters the market in 2006, it will provide approximately 35 percent more seats and approximately 50 percent more available floor space over the Boeing 747-400, the largest commercial liner expected to be in service at that time.[1] The initial version, A380-800 (Figure 5.1), is expected to have a capacity of 555 passengers in a three-class configuration and a range of about 8,000 nautical miles.

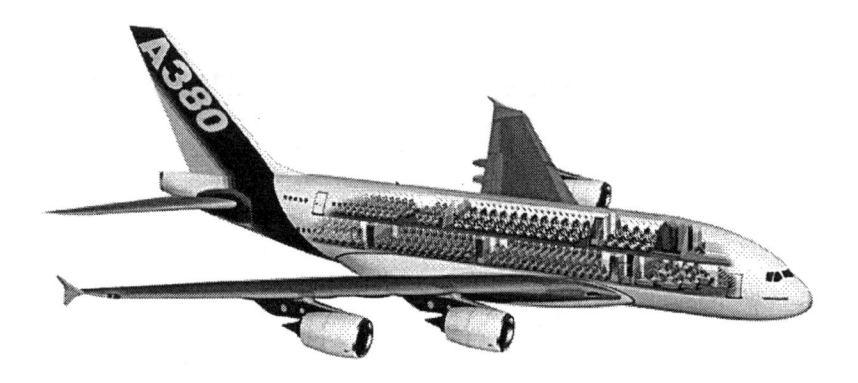

Figure 5.1 Airbus A380
Source: AIRBUS

Justification for this size of an airplane partly results from the assumption that the constrained infrastructure (the capacity of the airport and the air traffic control system) will not be able to accommodate the projected increase in global passenger air traffic and partly on the need to continue the trend to reduce the unit direct operating costs. This aircraft should ease ATC and runway congestion at major airports, especially in the Asia-Pacific region. Consequently, the A380 is envisioned to serve high-density trunk routes such as the ten airport-pairs (Figure 5.2) forecast to be the densest routes in the year 2019.

In certain high-density markets schedules are constrained by airport curfews, time zone differences, and connecting hub waves. Consequently, multiple aircraft are scheduled to depart within a very narrow window to accommodate all the traffic. Take, for example, the Hong Kong-London Heathrow market. The current schedule (for example, 12 November 2001) shows three airlines operating five daily flights, all leaving within a 55-minute interval (between 11:25 PM and 12:30 AM). British Airways has two departures with 747-400, one at 11:25 PM and the other at 11:35 PM. Cathay Pacific also has two flights operated with 747-400, one departing at 11:55 PM and the second departing at 12:30 AM. Finally, Virgin Atlantic has a flight with an Airbus A340 departing at 11:35

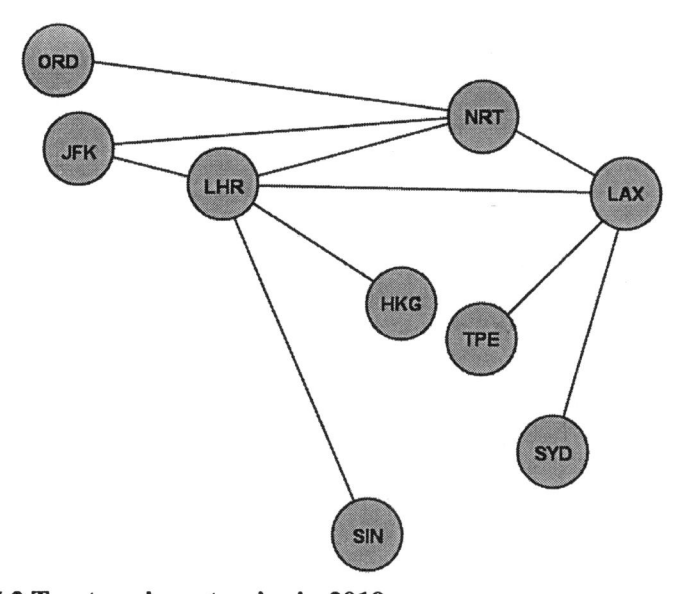

Figure 5.2 Top ten airport-pairs in 2019
Source: Based on AIRBUS, "Global Market Forecast 2000-2019," July 2000, p.39

PM. The average block time is about thirteen and one-half hours. Just as the five departures from Hong Kong take place within a 55-minute window, the expected arrivals at London's Heathrow are also between 5:05 and 6:20 AM—an equally narrow window of 75 minutes. This market is just one example of the list of markets shown in Figure 5.2 where the payload capacity and the lower unit operating costs could meet the needs of the marketplace and at the same time help to relieve ATC and airport congestion at both ends. On the other hand, all high-density markets (such as those shown in Figure 5.2) are not necessarily good candidates for high-capacity aircraft. Some high-density markets are driven by frequency, for example, JFK-LHR.

In addition to high-density point-to-point markets, the A380 is also expected to feed passenger traffic through large hubs partly to relieve runway congestion and partly to take advantage of its lower direct operating costs per seat—about 15 percent lower than the current long-haul Boeing 747-400. The market between London and Hong Kong has sufficient point-to-point traffic as well as some connecting traffic to benefit from a high-capacity aircraft. According to the data published by the United Kingdom Civil Aviation Authority, almost three-quarters of the passengers boarding aircraft in London for Hong Kong originated in London and only one-fourth arrived in London from other cities to make a connection in London. Similarly, of all the passengers arriving in Hong Kong from the flights out of London, about three-fourth are destined for Hong Kong itself and only about one-fourth make connections for other cities out of Hong Kong, such as Sydney, Perth, and Auckland. A high-capacity aircraft with good economics is ideally suitable for a market such as Hong Kong.[2]

There has been some market fragmentation on the North Atlantic (an increase in the point-to-point service with long-range aircraft of medium capacity),[3] and there could be some market fragmentation on the Pacific. However, there has also been consolidation within the airline industry evidenced by the development of dominant hubs, global networks, and strategic alliances. The consolidation process could indeed accelerate after the 11 September incidents. In any case, the fragmentation in the Asia Pacific region could be limited relative to the experience across the North Atlantic, partly because population tends to be concentrated within few large cities—for example, Seoul, Auckland, and Sydney, not to mention Singapore and Hong Kong—but also because income also tends to be concentrated within the same few large cities in countries in the Asia-Pacific region. However, neither population nor income is as concentrated in Europe or the United States, exemplified by the characteristics of Paris, London, New York, and Los Angeles.

The additional capacity of the A380 will offer airlines numerous possibilities for cabin configurations and passenger services in high-density markets. One possible configuration reported is 102 business class and 103 economy class seats on the upper deck and 22 first class and 328 economy class seats on the main deck. In one configuration, the seats would have a 2-2-2 arrangement in the first and business class sections and a 2-4-2 arrangement in the economy sections. (Figure 5.3) In a single class the aircraft is capable of transporting 850 passengers in truly high-density markets such as Tokyo-Osaka and Tokyo-Sapporo. In a two class configuration in the same high-density markets the capacity could be about 750.

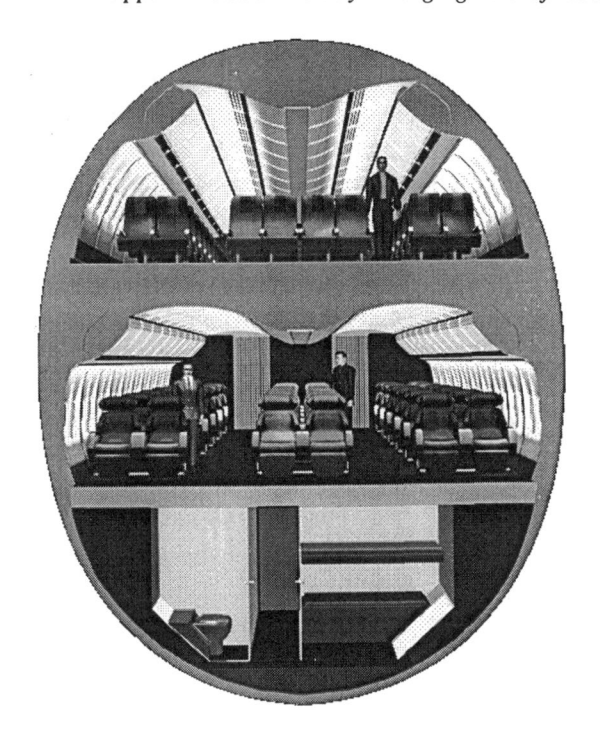

Figure 5.3 Cabin configuration possibilities
Source: AIRBUS

The additional space also provides an opportunity for greater comfort in the economy class by eliminating some of the health problems associated with long-haul flights in closely-spaced seats and in crowded cabins, such as the swelling of legs and the blockage of pulmonary arteries. Obviously, one option is to increase the seat pitch by a couple of inches. But a more desirable use of additional space might be to provide activity areas in the lower level where there is now sufficient room for passengers to stand-up straight.

In business class and first class cabins, the A380 can enable airlines to select from a broad spectrum of options for utilizing space in terms of luxury and privacy. It is interesting to note that three of the initial customers are Singapore Airlines, Emirates, and Virgin Atlantic—all three are well known for their service innovations. Here is just one example of a potential opportunity in long-haul flights. The lower deck has capacity for 13 pallets (or 38 LD3 containers). The forward pallet positions can be replaced to offer 12 bunk-bed positions for the crew. It is also possible to replace three forward pallet positions in the aft hold and offer 18 bunk beds. Moreover, certain pallet positions can be used for passenger amenities since the space provides a full two-meter standing height.

Technically, it is also possible to fit bunk beds in the containers themselves so that the decision can be made closer to the flight on the exact mix of containers carrying cargo and the containers with bunk beds for sleeping. However, these modular concepts have some drawbacks. For example, if the containers are light, they are too fragile and if they are sturdy, they are too heavy. There are also technical issues that need to be resolved with respect to the interface for air, heat, air conditioning as well as stairs. The best solution would be to adopt the pallet positions for crew or passenger rest areas on a semi-permanent basis. Consequently, these facilities can be removed during aircraft checks but not during normal turn arounds.

The press has reported a broad spectrum of potential amenities such as business centers, bars, saunas, showers, gymnasiums, shops, and restaurants. No doubt some airlines will use the space for value-added amenities such as sleeping beds and business centers. However, it is doubtful that these airlines will be willing to provide exotic amenities given that even a segment of business passengers is becoming price sensitive. In the 2000 International Air Transport Association Survey, 41 percent of respondents reported that their choice of airlines was restricted by their companies. This compared with about 25 percent in 1998.[4] Some airlines may use the two cabins simply to transport a large number of revenue-producing passengers, especially in the economy class, in a little more comfort. And, it is this second group of airlines that can deploy the high-density aircraft with its lower unit operating costs to make air transport affordable to increasing segments of the world's population.

The Boeing Sonic Cruiser Concept

The aviation industry has always pursued vigorously new technologies and innovations to provide higher speeds, capacities, and ranges. Subsonic commercial jets of the early sixties provided benefits in all three areas. Convair managed to squeeze a little more speed out of its 880 and 990 jets but at an enormous cost in fuel. Airlines decided, however, to sacrifice the small addition in speed and stay with the slower but more economical Boeing and Douglas fleets. The Concorde, on the other hand, did double the speed but it not only had high fuel costs but also a limited capacity and a limited range. Moreover, environmental constraints prohibited the aircraft from flying supersonic over populated regions. When the widebody aircraft entered service, they provided an increase in capacity and range as well as a reduction in unit operating cost. These attributes helped airlines to accommodate the growth of air travel. Even though these aircraft did not provide any significant increase in speed, their lower unit operating costs enabled airlines to lower fares and stimulates demand even further. In the past 20 years, for example, airlines achieved an almost 60 percent reduction in direct operating costs that enabled them to reduce passenger fares (in terms of yield) by an average of more than 3 percent per year. In subsequent years airlines, in fact, reduced the speed at which they flew the sub-sonic jets to achieve a further improvement in their economics.

Over two decades the dynamics of international markets have been changing significantly as a result of such factors as limited capacity of the infrastructure (air traffic control and runways), development of hub-and-spoke systems, availability of economical long-range, twin-engine aircraft with transoceanic capability, an increasingly liberalized regulatory environment, and a consolidation in the airline industry. As stated above, Airbus' interpretation of some of these forces was that ATC and runway congestion coupled with the need for economies of scale warranted an even larger aircraft for selected high-density markets. Boeing's interpretation of some of these forces, on the other hand, is a greater need for point-to-point services using smaller and in some cases faster aircraft, especially to meet the needs of premium-fare business travelers.

Based on its interpretation of the needs of the long-haul international markets, Boeing has proposed its new airplane concept, dubbed as the Sonic Cruiser (illustrated in Figure 5.4). It is envisioned to be a 200-250 seat aircraft capable of flying at about 95 percent the speed of sound (between 15 and 20 percent faster than today's standard subsonic jets) thereby avoiding the sonic boom problem. The Sonic Cruiser would be capable of flying at an altitude of 45,000 feet, and higher.[5] By incorporating such technologies as the double-delta wing, the aircraft should be able to comply with current and expected noise requirements.[6]

Higher-speed and higher-altitude operations can provide four benefits:

1. Flight time of trans-Atlantic trips can be reduced by about one hour and of trans-Pacific trips by up to two and one half hours.
2. In selected markets, aircraft productivity can be increased by flying round-trips in the same day.
3. Departure and arrival times can be changed in a few markets to capture additional demand.[7]
4. Flights at higher altitudes can provide smoother and more comfortable experience.

The viability of the projected performance targets depends primarily on viability of the engines. In other words, can engines be developed to enable the Sonic Cruiser with a capacity of about 250 to fly economically at Mach 0.95 or higher for distances up to 9,000 nautical miles? Second, would technology be available to enable the Sonic Cruiser to cruise at Mach 0.95 with a fuel consumption that is comparable to today's similar-capacity, long-range jets flying at between Mach 0.80 and 0.85? Rolls-Royce believes that technology to power the Sonic Cruiser with respect to these performance targets can be within reach during the time period envisioned for the development of this aircraft.[8] Moreover, even if one assumes that the Sonic Cruiser will consume 20-25 percent more fuel than today's jetliners, it is reported that it will have more than a 20 percent higher productivity because of its higher speed. For example, the Sonic Cruiser would be able to make a daily round-trip in some trans-Pacific markets that now require more than one traditional jet aircraft.[9] Los Angeles-Tokyo would be one example of such a market.

The economic viability of the Sonic Cruiser would depend on the total seat-mile costs and the need, if any, to charge premium fares over the current the current first and business class fares. With no or a small fare premiums, the concept makes sense in those markets where there is sufficient first- and business-class traffic on a year-round basis. Examples

Figure 5.4 Boeing's proposed Sonic Cruiser
Source: Boeing Commercial Airplanes

include New York-London, New York-Tokyo, London-Singapore, and Singapore-Sydney. The degree of market penetration will depend on the level of fare premium and the amount of time saved. During February-May 1993, a research group conducted discussions with 40 focus groups involving corporate travel decision makers in Los Angeles, Tokyo, and New York. The purpose of the research was to determine the interest in high speed commercial transport with respect to the time saved in travel. One conclusion was that there would need to be a minimum of two and one-half hours of time savings for passengers to be willing to pay a 20 percent premium over subsonic fares. Consequently, it would appear that the Sonic Cruiser could obtain such a premium from passengers in those trans-Pacific markets where it could produce a minimum time savings of two and one-half hours.[10]

Regional Markets

In the 1950s, the Boeing 707 and the Douglas DC-8 started the jet revolution in the aviation industry worldwide. In the 1990s, the Bombardier CRJ-200, the Embraer ERJ-145, and the Fairchild Dornier 328JET produced a similar revolution. These aircraft, developed to serve short and medium-haul markets, were preferred by passengers to turboprops because they were fast, quiet, and comfortable. The airlines preferred them because they were highly productive, reasonably affordable, and had competitive operating costs in regional markets. The FAA changes in its regulatory policy also played a crucial role in the tremendous popularity of the regional jets. The FAA not only required regional airlines to install sophisticated cockpit safety systems in smaller airplanes, but also the agency mandated in 1997 that regional airlines operate under its more stringent—and more costly—Part 121 rules, previously applying only to the mainline carriers.

There is no clear definition of regional markets. These markets tend to be under 1,000 miles in length of haul (under two hours in flight time), with thin density, usually served with aircraft less than 70 seats, and usually connecting two secondary communities through hub-and-spoke systems. Prior to the entry of specifically designed regional jets at the beginning of the 1990s in the USA, these markets were served with turboprops aircraft or did not have service. With the introduction of jets, regional markets experienced explosive growth in the past ten years based partly on the competitive economics of these aircraft on partly on their passenger appeal (based on such considerations as noise and modernity). However, despite the greater appeal of regional jets some turboprop aircraft not only contain high technology but they provide benefits of lower unit costs and therefore are more suitable for selected markets.

Most regional jets are flying in North America. Initially, they worked the hub-and-spoke environment. See Figure 5.5. In one set of markets, airline brought in connecting passengers from longer spokes. In a second set of markets, airlines substituted turboprops for regional jets. In a third set, airlines substituted mainline jets for regional jets. In the fourth set, they complemented the turboprops. And, in a fifth set, airlines complemented the mainline jets with regional jets. Beginning in 1998, airlines began to increase the use of regional jets to serve point-to-point small and medium size communities from major metropolitan areas such as Boston, New York, and Washington.

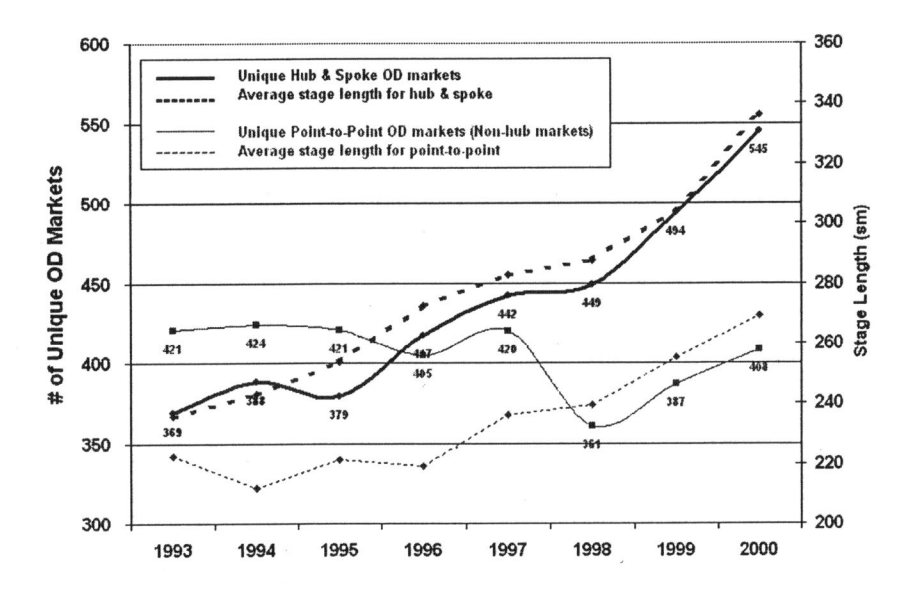

Figure 5.5 Penetration of regional jets in hub-and-spoke and point-to-point markets
Source: Bombardier Aerospace

Service could also increase from the smaller but less congested airports located near the larger metropolitan areas. Examples include Long Beach, California (near Los Angeles), Worcester, Massachusetts (near Boston), and Gary, Indiana (near Chicago). The future success of regional aircraft depends on a number of factors:

1. Labor issues relating to the pay inequity between pilots flying regional aircraft and those flying mainline aircraft,
2. Potential implementation of economic market-based solutions to airport congestion, and

3. Potential slot allocation regulations in Europe to reduce the number of available slots at selected airports.

The primary hurdle facing the even greater use of regional jets within US markets is the existence of "scope clauses" written into US pilots contracts. These clauses limit the degree to which major airlines can substitute their flights with larger jets with the flights of their regional partners who are using the regional jets. These scope clauses protect the pilots flying mainline jets from management outsourcing the mainline pilot jobs to the pilots flying for the lower-cost regional airlines. However, the growing role of regional jets has led the pilots of regional airlines to seek higher wages, an outcome that could redefine the economics of regional jets.

The application of regional jets is markedly different in Europe. First, they are being used primarily to serve point-to-point secondary cities by bypassing congested hubs. Second, the average size of regional aircraft in Europe is higher then that in the United States: 67 compared to 31.8 in the United States. This difference exists because regional operators have a larger percent of jets in their fleet—49 percent compared to 25 percent in the United States. Third, unlike in the US, a number of European regional carriers fly mainline narrow-body jets such as the MD-80 and A319 in either scheduled or charter services.[11] Fourth, to make the service somewhat similar to the mainline, European regional carriers are using larger aircraft. And finally, scope clauses do not exist in the pilot contracts of major European airlines with regional partners.

The regional jet manufacturers are producing and have plans to produce regional jets with capacities up to 90 seats. Larger regional jets have better economics. Their costs do not increase in proportion to the number of seats. Therefore, larger jets offer lower unit operating costs. Second, some of the larger aircraft such as the Fairchild Dornier 728JET can provide some flexibility with respect to cabin configuration due to the larger cross section of their fuselage. The 728JET could set a new standard in passenger space and comfort for single-aisle airliners for regional markets. The overall width of the 728JET (128 inches) can easily accommodate 5 (18 inch width) seats in a row. See Figure 5.6. An airline could also offer a premium-class cabin with 4-abreast seats. Consequently, there is a real opportunity for regional jets, particularly, for aircraft such as the 728JET with a wide cabin to provide the comfort levels of narrow-body jets but that are much more economical than the early versions of mainline airliners such as the DC-9 and the 737s.

In the after 11 September world, the regional jets are expected to play an even more significant role. They could replace mainline jets on many more routes to adapt to the lower levels of passenger traffic, constrained only by scope clauses and available financing. The replacement of the older larger mainline jets (such as Boeing 727s and 737s and the Douglas DC-9s) with the new regional jets not only provides economic benefits in terms of greater harmonization between capacity and demand but they are also economical with respect to maintenance and fuel costs. Second, they could also be used to provide nonstop service directly between selected spokes. The passenger data shows, for example, that in the year 2000, there were 20 cities that had more than 1000 daily passengers who traveled between 200 and 1000 miles using connecting services.[12] Examples include Orlando, Indianapolis, Kansas City, Tampa, and Washington, D.C. However, one cannot overlook the fact that part of the reason the regional jets are economical is because their economics are based on paying pilots lower salaries. With comparable salaries, the economics look a lot different.

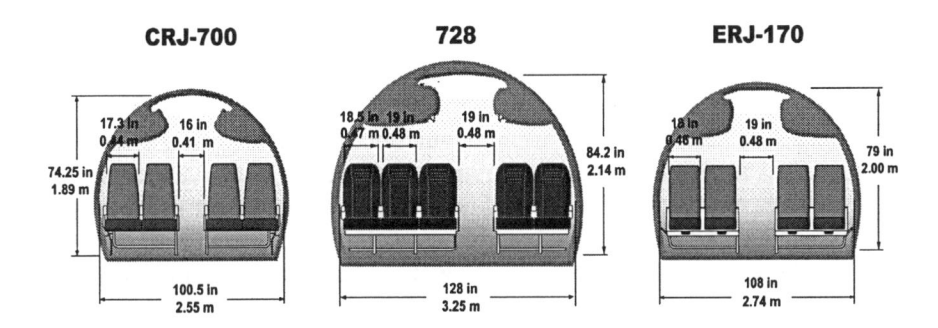

Figure 5.6 Cross sections of jets for regional markets
Source: Fairchild Dornier

Personal Transportation

Strong interest in personal transportation in private aircraft has been fueled by seven major factors: (1) increasing levels of delays due to congested airports and air traffic control systems; (2) increase in the hub-and-spoke activity of major airlines; (3) very high fares for travel involving no restrictions; (4) deteriorating levels of customer service; (5) increasingly diversified and geographically dispersed economy; (6) decreasing levels of privacy, and (7) the development and introduction of a broad spectrum of new business jet models. Now we can add an eighth factor, the extra time needed to process passengers at major airports in light of the new security measures after the 11 September events and extra security requirements of high-level people. Let us take the combined impact of the first two factors mentioned. According to analyses conducted by NASA, within the United States, for trips under 500 miles the average speed (from the time a person leaves his or her home or office and arrives at the destination) is only 50 or 60 miles per hour—not much higher than driving a car.[13] And the situation has become worse after the 11 September incidents. The fifth factor relates to the trend that the U.S. economy has been becoming more dispersed as businesses relocate from urban areas to rural communities.

To overcome inconvenience of air travel, poor customer service, and in some cases the high cost of last-minute travel, passengers have looked to the use of private aircraft for personal transportation. But, until recently personal air transportation has not been economically viable other than for a very small segment of the population. However, partly as a result of advanced technology, partly as a result of new business models, and partly as a result of the U.S. government's decision to invest more money in R&D related to aviation technology as well as the decision to open up literally thousands of smaller underutilized airports, personal transportation by air is increasingly becoming technically feasible and economically viable. The first part of this section provides an example of a new business model—fractional aircraft ownership. The second part provides two examples of innovative aircraft technology that has the potential to bring about major changes in how passengers can travel using personal aircraft.

Fractional Aircraft Ownership

Private aircraft enable individuals and companies to travel more efficiently by controlling their own schedules and by flying to and from airports located closer to their points of origin and destinations. In addition, private aircraft provide individualized service and privacy. This method saves passengers considerable time over using commercial services. Passengers save time by flying directly to their destinations and gain productivity since they can work en route. Convenience occurs because the passenger determines the aircraft's schedule rather than the commercial airline dictating the passenger's schedule. The passenger can travel directly between points of origin and destination, avoid congested airports and terminals, and avoid considerable air traffic delays at major airports. Moreover, the use of private aircraft provides privacy with respect to the sensitivity of the information discussed on board the aircraft. However, private aircraft are expensive to own, operate, and maintain.

The fractional aircraft ownership program gives the benefits of owning a private aircraft at a fraction of the cost of owning the whole aircraft. The program not only eliminates the responsibility to manage the aircraft, it makes it possible for the customer to tailor the size of the investment to meet the customer's transportation requirements—the number of hours flown each year, trip lengths, and number of people in the group. The management company manages the aircraft (with respect to crew, maintenance, and so forth) and provides other operational support such as in-flight meals, ground transportation and accommodations.

With fractional ownership interest, the customer has access to a large fleet of private aircraft. For example, not only does the fractional ownership owner have access to all aircraft similar to the one purchased by the owner but also potential access to even larger aircraft through an interchange agreement. By paying additional charges, such agreements enable owners to select aircraft types different from those that they own to satisfy specific trip requirements. The fractional owner is not guaranteed access to a specific aircraft, only that the owner will have access to an aircraft of the same make or better. The bigger the size of the management company the shorter the amount of notice necessary to schedule a flight.

Large management companies such as NetJets (a division of Executive Jets that itself is a Berkshire Hathaway company) can provide an airplane with as little as for four hours' notice.

According to NetJets, fractional ownership is cost effective for individuals and businesses that travel between 50 and 400 hours per year with departures from multiple airports. For individuals flying less than 50 hours a year it is more cost effective to charter an aircraft instead of acquiring a fractional ownership. For individuals flying more than 400 hours per year it is cost-effective to own an entire aircraft. However, fractional ownership may still make economic sense for individuals flying more than 400 hours pre year if the trips vary in length and originate from multiple locations. Moreover, fractional ownership is particularly valuable if the business needs more than one aircraft on a given day.

The smallest fraction with NetJets is 1/16th that provides 50 hours of flight per year. At this level of ownership interest the customer has up to 50 hours of flight time available per year. The commitment is generally made for five years. The monthly management fee covers such costs as pilots, insurance, hangar, dispatching, meteorology, catering and administration. The hourly costs pay for fuel, aircraft maintenance, and landing fees. The hourly costs are fixed and predictable. Moreover there may be tax advantages to fractional ownership related to the depreciation of the asset.

For a one-sixteenth share, the ownership cost would vary from around US$400,000 for a new aircraft such as the Citation V Ultra to about US$2.7 million for an aircraft such as the Gulfstream V. Monthly management fee would vary between around $5,000 and $17,000 for the two aircraft mentioned. Finally, the cost of each flight hour for the same two aircraft would be $1,300 for the Citation and $3,000 for the Gulfstream.[14] In 2001, NetJets operated a fleet of about 300 business jets and had on order another 300 aircraft. The company is extending the range of aircraft it manages by including the Boeing Business Jet. The company has about 2,400 fractional owners. About 30 percent of the owners are individuals and about 70 percent are corporations.[15]

The fractional aircraft ownership business has been increasing at a significant rate since the beginning of the 1990s on the basis of convenience, efficiency and productivity. According to the US National Business Aircraft Association, in 1993 there were 110 companies or individuals who owned a share in business aircraft. In the year 2000, the number had climbed to 3,694.[16] Since 11 September, the interest in fractional ownership has grown enormously partly because of the convenience and partly because of security. Mainline airlines can require passengers to come to the airport up to three hours prior to departure, but owners of private aircraft can arrive up to within 10-15 minutes of their intended departure time. In the United States, scheduled airlines operate out of approximately 580 airports, but owners of private aircraft can fly in and out of approximately 5,400 airports.[17]

Partly because this segment of the aviation business has been growing at double digits for the past five years, partly to reduce the potential defection of some first and business class passengers, and partly to meet the needs of important corporate customers, a number of commercial airlines have begun to examine different ways of entering this segment of the business. British Airways reached an agreement with a very large corporate aircraft charter broker to provide aircraft to passengers arriving on British Airways. Flights would transport them to destinations either not served by British Airways or destinations with no scheduled air service. Air Canada has entered into a similar arrangement with a corporate jet operator only, unlike British Airways' partnership, Air Canada has equity in its strategic partner.[18]

In the United States, some commercial airlines have also evaluated potential opportunities that may add value by providing synergies with their existing services, exemplified by UAL Corporation's (parent of United Airlines) decision to enter into this business. UAL took options on 135 Falcons and Gulfstreams and indicated the possibility of buying smaller aircraft as well with the ultimate size of the fleet to be between 200 and 250 by the year 2005. With the use of business jets, a commercial airline can presumably attract and retain high-yield passenger traffic through the management of fractionally owned aircraft providing on-demand services, corporate shuttles, and scheduled feeder services linked to their mainline international air transportation services. The corporate shuttles are envisioned to provide air transportation services for specific corporations between specific cities. One hurdle that a commercial airline could face would be significant conditions established by organized labor that would limit the company's formation, ownership, and operation of such a business entity.

Business opportunities for fractional ownership are so significant that aircraft manufacturers, management companies, and government agencies have even considered the possibility of developing a supersonic business jet with a purchase price approaching US$100 million. The total market for the supersonic business jets is estimated to be about 250 aircraft. NetJets forecasts that there is a market for about 50 fractionally owned supersonic business jets. The majority of these are likely to be owned by corporations with a few owned by extremely wealthy individuals, governments (for the purpose of evacuations and diplomatic missions), and Medevac (for transporting organs). Customers will pay a higher price for greater speed. NetJets substantiates this claim by pointing to the high demand for the Cessna Citation X business jets in its fleet that enable their owners to travel from the US West Coast to the East Coast in four hours and return in four and a half hours. According to a NetJets survey of its long-time owners, there are customers willing to buy a 1/8th share of a supersonic business jet at US$10 million with $4,000 per hour operating costs if such an aircraft could fly at speed of Mach 1.8 and have a trans-Pacific range (5, 000 nautical miles). Customers would accept a cabin the size of the current Citation X for trips less than five hours.[19] The major challenge appears to be not the economics but rather the technology challenge with respect to the sonic boom, engine noise, and emissions.

Personal Aircraft

At the other end of the spectrum from the supersonic business jet is the Eclipse 500, a new super-light twin-turbofan business jet expected to cost under a million dollars—a fraction of the price of even low-end, entry-level business jet such as the Learjet 31A that costs between 6 and 7 million dollars. The Eclipse 500—Figure 5.7—is expected to be in service in 2004. The aircraft is reported to have a capacity of six (including crew) and a range approaching 1,300 nautical miles to serve a market such as New York-Miami. The commercial success of the Eclipse 500 is based on the expected breakthroughs in: (a) the new small engine technology (based on the innovative high thrust to weight engines developed by Williams International); (b) innovative manufacturing techniques (such as "friction stir welding" for fabricating the primary structures);[20] (c) digital information technology; and (d) high-technology systems integration techniques. Such an aircraft could easily enable individuals in the $100,000+ income group to participate in the fractional ownership programs. And since the aircraft is expected to have the capability of using 2,500-foot runways, these individuals could use thousands of airports within the USA instead of the 580 or so used by commercial airlines.

Taxi service is another potential innovative use of the Eclipse 500. The Nimbus Group has already ordered 1000 Eclipse aircraft to create a nation-wide on-demand air taxi service.[21] Service is envisioned to be available at about US$500 per hour with no dead-head fees. Such a price could generate an enormous demand from the middle management levels at Fortune 500 companies. Moreover, the availability of an Internet-based reservation system could easily generate additional passengers who could share a taxi service and divide the $500 per hour cost among three or four passengers. It is reported that the Nimbus Group is considering operating the taxi service under the FAA's Part 135 air taxi operating certificate as well as establishing a franchise system encompassing pilot-operators flying company-owned aircraft.[22] These two examples illustrate the degree to which an aircraft such as the Eclipse 500 could open business aviation to a new era in air transportation.

The component of the general aviation industry designing and producing single-engine aircraft had not introduced any new innovative designs in the past 20 years, based partly on the decrease in demand since the late 1970s. In the mid-1990s, the Cirrus Design Corporation took on the challenge of designing and producing a single-engine general aviation aircraft based on a platform of new technology. The primary market for the aircraft was envisioned as the potential pilot-owner segment for which safety was the top most requirements. Secondary considerations were price, operating costs, and comfort.

Figure 5.7 Eclipse 500
Source: Eclipse Aviation

The Cirrus Design Corporation designed its SR20 to meet all these requirements—an aircraft that entered service in July 1999. Examples of technology incorporated to make the aircraft extremely safe include, (a) "wing cuffs" that can make it easier for the aircraft to recover from a stall, and (b) a parachute system to descent the entire aircraft safely on the ground in life-threatening situations such as engine-failure, fuel starvation, and critical structure failure. See Figure 5.8. In addition to the parachute, the aircraft has the necessary padding to avoid a hard landing. The aircraft is also equipped with the latest-technology avionics such as a GPS navigation system and advanced engine instruments.

Constructed with composite materials, the SR20 is designed to cruise at 160 knots and with a range of about 800 nautical miles. With full fuel, the aircraft has a payload of more than 600 pounds. The economy of the SR20 can be gauged from the fact that the aircraft can cruise at 160 knots on just 10 gallons per hour.[23] The SR20 is a remarkable aircraft that offers safety, comfort, and competitive economics for enthusiasts to fly themselves from point A to point B at their own schedule.

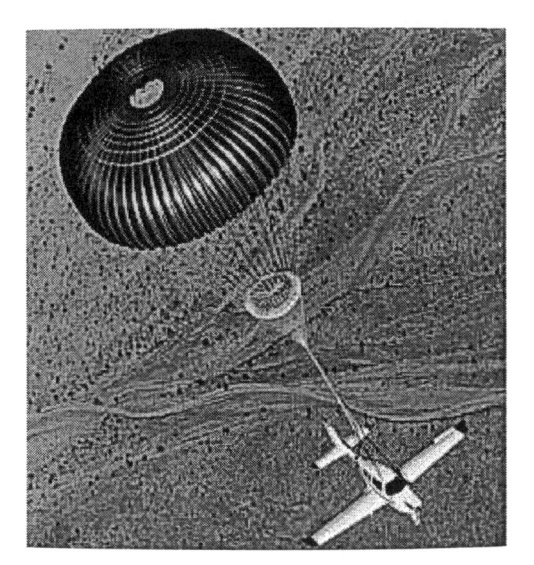

Figure 5.8 Cirrus SR-20
Source: Cirrus Design Website

Advanced technology aircraft, such as the SR20 and the Eclipse, have the potential to make air transportation not only truly "personal," but also convenient and affordable. Coupled with the other high-technology aircraft described in this chapter, the broad spectrum of emerging aircraft has the potential to redefine air transportation.

Notes

[1] *Air International*, July 2001, p.36.
[2] The United Kingdom Civil Aviation Authority, "1998 CAA Survey CAP 703," November 1999.
[3] Boeing Commercial Airplanes, *Current Market Outlook*, June 2001, p.42.
[4] The International Air Transport Association Corporate Air Travel Surveys, 2000 and 1998.
[5] *Airways*, July 2001, p.56.
[6] *Air Transport World*, August 2001, p.37.
[7] *Aviation Strategy*, July/August 2001, p.4.
[8] *Airways*, July 2001, p.56.

[9] *Airways*, July 2001, p.56.
[10] Research conducted by Plog Research, Inc for the High Speed Commercial Transport International Working Group, February-May 1993 and reported in Adam Brown, "Size or Speed," A presentation at the ISTAT 8th Annual European Conference, Venice, Italy, 16 October 2001.
[11] *Air Transport World*, October 2001, p.87.
[12] Analysis conducted by the North American Division of Atraxis.
[13] Goldin, Daniel, Keynote Address at the Aircraft Owners and Pilots Association Convention, Palm Springs, California, 24 October 1998.
[14] NetJets brochure showing the typical charges for the year 2001.
[15] *Airline Business*, August 2001, p.78.
[16] *Financial Times*, 6 November 2001, p.12.
[17] *Financial Times*, p.12.
[18] *Airline Business*, August 2001, p.76.
[19] Interview with Richard G.Smith, Columbus, Ohio, 22 October 2001.
[20] Collogan, Dave, "Eclipse Changes Relationship with Williams, Schedules for New Twin Turbofan," *Business Aviation*, 23 July 2001.
[21] *Business and Commercial Aviation*, a Report by David Rimmer 1 October 2001.
[22] *Flight International*, 25 September 2001.
[23] Hingdom, David "Plastic Planes, Part Two: The Cirrus SR20," A report available from http://www..avweb.com/articles/cirrussr20/, p.8.

Chapter 6

Forces Transforming the Air Cargo Market

The air cargo segment of the aviation industry has both grown and changed at a faster rate than the passenger component. The air cargo component has been adapting to the evolving needs of shippers and capitalizing on (a) the growing segments of the marketplace, and (b) emerging technology. This applies to both aircraft and information management. This chapter highlights the key transformative forces—including the interaction among them—shaping the air cargo market.

Changing Needs of Shippers

Globalization—integration of world economies and reduction of protectionist policies—is proving to be a double-edge sword for the air cargo industry, which has increased international trade and led to a relocation of manufacturing facilities. Although the average annual growth rate of the international express carriers—also known as integrators—has been declining, from 20 percent until the mid-1990s to 18 percent since then, it is still expected to grow at an annual rate of about 13 percent per year for the next twenty years. The average weight of international shipments has been increasing from 6 pounds in 1992 to 10 pounds in 1999.[1] Globalization has encouraged manufacturing facilities to be located where costs are low and value added is high.

Globalization has increased competition both outside the air cargo industry as well as within the air cargo industry. Outside the air cargo industry, competition has forced all industries to compress their cycle times. This trend benefited the air cargo industry, helping to reduce delivery times and inventories. Competition has increased within the air cargo industry not only among the scheduled airlines but also among various components of the air cargo industry—basically between integrated and non-integrated carriers to meet the changing needs of shippers. This increase in competition is partly responsible for the continuous decline in cargo rates. See Figure 6.1.

In addition to globalization, more and more production is now consumer driven, requiring shippers to align their supply chain with the customer chain. To meet

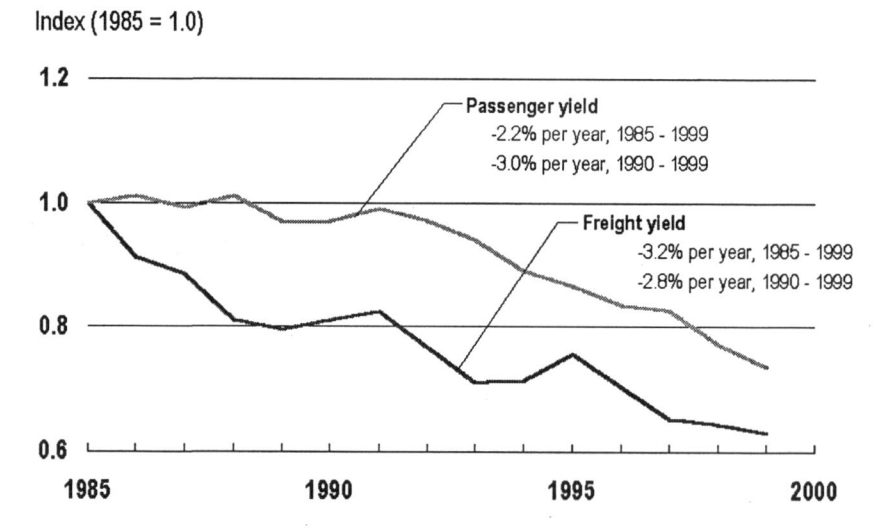

Index (1985 = 1.0)

Passenger yield
-2.2% per year, 1985 - 1999
-3.0% per year, 1990 - 1999

Freight yield
-3.2% per year, 1985 - 1999
-2.8% per year, 1990 - 1999

Figure 6.1 Passenger and freight yield trends
Source: Boeing Commercial Airplanes

this requirement, shippers want the non-integrated carriers to provide similar services to those provided by the integrated carriers (who own or control different parts of the transportation product—aircraft, trucks, and information systems) who have invested heavily in information technology. Shippers are not willing to pay the non-integrated carriers the higher rates charged by integrated carriers even though they want those carriers to match the capability of integrated carriers. While shippers demand high level of service, they also want it at a lower price. [2]

Shippers want a broad spectrum of products, reduction in transit times, flexibility in services, and online accessibility. They seek time-definite services that relate to guaranteed delivery times such as the express, the next day, and deferred which generally implies two or more days. Such services provide shippers with the option of choosing shorter delivery times for higher rates or longer delivery times for lower rates. The need for shorter transit times is obvious from the fact that shipments in general spend only about 10-15 percent of their total transit time in the air and 85-90 percent on the ground, both going to and from airports as well as time spent at the airport prior to departure and after arrival. Consequently, total transit time for an international shipment can be four to six days. The proliferation of shippers' interest in time-definite services can also be observed from more and more shippers within the business-to-business market (expected to be four to five times the size of the business-to-consumer market) showing an interest in the type of services provided by integrated carriers—services that previously have been used to a large extent by the business-to-consumer segment. Encouraged by freight forwarders to expand the time-definite services, passenger carriers are beginning to notice the demand for such services. For example, Lufthansa Cargo, who is reported to have pioneered the adoption of time-definite freight services, now, derives more than two-thirds of its revenues from providing time-definite services.[3]

Flexibility refers to the ability of the shipper to choose the routing, type of space (belly vs. freighter), and type of handling (door-to-door service vs. airport-to-airport transportation). In addition, some shippers may want to conduct transport transactions manually while others may want to do it thorough the Internet. Some may not want to go through Web-based auctions and virtual transportation exchanges. They may prefer to deal with transport operators because real companies can take responsibilities for the delivery of freight. Some shippers may prefer to use auction groups to obtain a break on rates.

Finally, some shippers may want online accessibility not only to obtain information and make bookings but also to receive information on the recovery process—for example, tracing and tracking. Shippers of high-value consignments demand guaranteed on-time service plus tracking and tracing capability in real time. These standards can only be met if there is seamless integration in the transportation chain. They also want to be able to track shipments using a mobile phone and be able to measure performance against service levels promised to them. Shippers have two problems with online accessibility, mostly with the non-integrated carriers. First, some airlines have their own systems that are not compatible with the customers' own systems. The root of this problem appears to be the difficulty involved in making the legacy systems of airlines compatible with the Internet. See Chapter 7. The second problem is the lack of common industry standards based on customer needs.

Shippers' needs have been evolving. Changes mentioned here include the globalization process and consumer purchasing behavior. Initially, shippers used air transportation because of speed, ability to handle fragile consignments, higher security, and lighter packaging—valuable attributes for shippers of high-value products and perishable products. In the past two decades, shippers have been seeking more and more logistics support that minimizes transportation costs and cuts distribution costs. More recently, some shippers realize that in the increasingly competitive environment, a few sophisticated cargo service providers can assist them with their global logistics strategies.

Recently, eleven large non-integrated air cargo carriers responded to a survey to determine their interpretation of shipper requirements over the next five years. They ranked 13 areas of service in terms of importance. On a scale of one to five (one being unimportant and five being very important), the top two service elements (tied with a near-unanimous ranking of 4.9) were "shipment tracing and tracking" and "information technology." The next two areas of importance (also tied with a ranking of 4.5) were "operational capabilities" and "information reporting capabilities." The next two elements (again tied with a ranking of 4.4) were "expedited services" and "high-technology warehouses."[4] It is interesting to note how shipper requirements may evolve over the next five years. Some requirements have not changed. "Shipment tracing and tracking" and "operational capability" rank high in importance now and are expected to continue as top priorities in the future. However, even though the ranking of "proven expertise/reputation" slipped from number two in the current environment to number eight in the future environment, the weight received by this element actually increased from a 4.1 to a 4.3. Similarly, the ranking of the service element "high-technology warehouse" improved from number 13 in the current environment to number six in the new environment. Its actual weight also increased from 3.0 to 4.6.[5]

The survey does not mention security because it's been assumed that cargo is more secure when it moves by air than by other modes of transportation. However, after the 11 September incidents, not only are the costs of security likely to increase but the transit times may worsen. Moreover, even if shippers are willing to work around the increase in transit times, the more severe problem is the unreliability and uncertainty of delivery times for businesses that depend on known times of inventory arrival. Customs clearance has always been a problem partly because of increased transit times and partly because of the uncertainty of delivery times. This problem worsened after 11 September. One advantage has been that documents have not needed customs clearance to the same degree as other shipments. Now, both types of service providers are likely to experience increased and varying delays by shipments waiting to be cleared.

Shipper requirements vary significantly by region. For example, unlike within the United States where integrated carriers such as FedEx and UPS make domestic deliveries, within Europe these deliveries tend to be made by ground-based parcel operators. Of the large four integrators, TNT has had a significant presence in the local delivery market within Europe. Second, although global service providers have established their websites and the European Union is a single market, by and large, because of national preferences and languages. Shippers in some European countries prefer to buy off their own national websites. Cross-border e-commerce is fairly limited.[6]

Recently, some shippers saw that in the increasingly competitive environment, a few sophisticated cargo service providers can assist them with their global logistics strategies. Historically, shipping costs have been about one half of the total logistics costs. Coordinated efforts of shippers and air cargo service providers have helped businesses to reduce their total distribution costs by the use of air cargo. However, to make further inroads in the reduction of distribution costs, shippers need to identify not only their service requirements but also their requirement patterns, distribution of their shipments (number of pieces and weight), distribution of destinations and the distribution of the shipments by the level of urgency. Learning the traffic patterns can save a lot of money. Carriers may be able to offer even lower rates if shippers can lower their costs of doing business and provide long-term partnerships. Moreover, even if a carrier cannot reduce its rates, it might provide value-added services such as electronic billing—a process that may save the shipper significant costs in the auditing process and, in addition, help track shipment patterns. Making it easier for an airline to work with the shipper forms one side of the story. It is important for the shipper to be able to work more easily with the airline. Not too long ago a freight forwarder needed to contact multiple airlines to get quotes for transporting individual shipments, an extremely time consuming process. Shippers, direct or indirect—freight forwarders—need to be able to enter the detail of the shipment (weight, type, and so forth) and the required level of service (delivery time, freighter or belly space, and so forth) and receive quotes quickly from interested carriers. Such services are just beginning to be available in the marketplace.[7]

Observing the branding activities of large non-integrated carriers shows that the leading air cargo service providers have a good handle on the shipper requirements—reliability, operational capabilities, and worldwide coverage.[8] However, offering shippers these service elements marks only one part of the profit equation—the strategy to segment the marketplace based on product and shipper profitability is another part.

Evolving Industry Structure

Independent Units of Combination Carriers

Despite the high growth of integrated carriers and all-cargo carriers, one half of air cargo is still transported in the bellies of passenger aircraft. Take, for example, Lufthansa—the world's largest cargo carrier in terms of ton-miles. About 40 percent of the cargo transported by Lufthansa Cargo rides in the bellies of Lufthansa's passenger aircraft.[9] However, the percentage of cargo transported in freighters has been increasing within the industry. The rate would have increased even more if passenger airlines had created truly separate cargo divisions.

In recent years, a few large passenger carriers—such as Lufthansa and Singapore—have created truly separate cargo divisions to make their own decisions. They are separate in the sense that they manage their own business risk. At the other end of the spectrum are some passenger carriers who have separate cargo divisions that are nothing more than units that simply sell the belly space of their passenger aircraft. They have virtually no input in the strategic direction of their units or their parents. They take costs as given and they are usually only concerned with generating whatever revenue they can. They tend to rely on incremental pricing, making low prices to attract shippers. Such a philosophy does not encourage the investment in innovative product development, either aircraft or infrastructure and in turn can reinforce the need to implement marginal-cost pricing strategies. And the situation has become worse by the availability of new wide-body passenger aircraft with large belly capacities.

Truly separate divisions actually buy the belly space from the passenger divisions of their parents and then try to sell it at profitable rates. Few leading carriers even buy belly space from carriers other than their own parent. Lufthansa Cargo, for example, reported that it has the responsibility for selling belly capacity of not only Lufthansa but also Lufthansa CityLine, Condor Flugdienst and Spanair.[10] Pricing belly space is a difficult process. First, allocating costs must be realistic. Separating the costs of freighter operations is not difficult but cost allocations for freight carried on passenger flights is another story. It involves allocations of aircraft, crew, maintenance, and fuel costs as well as other expenses such as landing fees. This is just the beginning of the cost allocation process. True allocations must also include some payment by the cargo division for traffic rights and airport landing slots. In addition to realistic allocations of costs, truly separate units are also required to be more responsible financially and more nimble, that is, quick to react to developments in the marketplace. Unlike the passenger side where demand can be easily stimulated by promotions and lower prices, it is much more difficult on the cargo side where the demand depends on other factors such as industrial production, trade, and retail sales.

Consolidation and Blurring Boundaries

One of the most significant trends in the air cargo industry is the consolidation process that encompasses mergers and acquisitions (within the freight forwarder segment), strategic alliances (within the airline industry), and blurring of boundaries among the different categories of air cargo service providers. Merger activity is exemplified in the freight forwarder segment, for example, by the fact that the top 15 freight forwarders now account for about 50 percent of the total tonnage worldwide.[11] The primary forces behind the consolidation/blurring boundaries process are the (1) higher growth of cargo traffic relative to passenger traffic, (2) evolving requirements of shippers, (3) phenomenal growth of integrator carriers, (4) reduction in the rate of growth in the overnight envelope traffic, (5) forthcoming deregulation of postal services within Europe, (6) increasing level of competition within the industry (within the air transportation mode as well as among different modes of transportation), and (7) availability of technology—aircraft, the Internet, wireless communications, and so forth.

A number of passenger carriers decided to refocus on cargo traffic because of declining passenger yields, increase in the value of products shipped by air, and the emergence of new markets as a result of globalization of world economies. Different carriers took different approaches to focus on the air cargo market. Some passenger carriers decided to test the market by "wet-leasing" freighters to provide service in long-haul intercontinental markets. The "wet-lease" or ACMI airlines are cargo carriers who provide freighter service for traditional airlines that includes not only aircraft but also crew, maintenance, and insurance. Examples include Atlas, Evergreen, Gemini, and World. During the 1990s, service provided by the ACMI carriers increased at an average annual rate of 21 percent.[12] Such contracted services have proven to be valuable for the traditional airlines during seasonal periods. The traditional carriers have also benefited from the flexibility provided by the ACMI carriers by enabling the former group to test market response to freighter service.[13]

Formation of strategic alliances is an example of a different strategy followed by passenger carriers. The idea behind forming an alliance is to form a global logistics service by coordinating route networks and schedules so that market share can be increased in a larger number of origin-destination markets. This objective can be achieved by linking additional spokes through the major hubs. Movement of freight through hubs is more concentrated than the movement of passengers through hubs. For example, the top ten passenger hubs account for about one-third of passenger movements whereas the top ten cargo hubs account for about two-thirds of all the cargo movements. Examples of the three top cargo hubs are Memphis and Los Angeles in the United States, and Hong Kong, and Tokyo.

Lufthansa Cargo, Singapore Airlines Cargo, and SAS Cargo joined forces in 2000 to form New Global Cargo, an organization in which Star alliance members can cooperate to develop time-definite products. This alliance should do well given that about 70 percent of the total air cargo is flown between or within the three continents—North America, Europe, and Asia.[14] Lufthansa Cargo and Singapore Airlines cargo are enormous in size, being number one and the number three in the world in terms of ton-miles. They also complement their networks. Although SAS Cargo is only a little more than 10 percent of the size of Lufthansa Cargo or Singapore Cargo in terms of revenue ton-miles, it has a similar philosophy. It is a separate corporate entity like the other two partners. Second, it is willing to adopt the principles of value-based management—discussed in Chapter 7—exemplified by the integration of their sales teams. Furthermore, Scandinavia is also the home of some high-value shippers like Nokia, Ericsson, Volvo, and Farmacia. This alliance's market potential could increase enormously if a large carrier from North America were to become a partner and adopt a similar management philosophy.

Although the strategic alliance concept among passenger airlines is spreading to the cargo side, it remains to be seen whether the degree of coordination on the passenger side can be carried out on the cargo side. It could work for cargo units that are truly separate business units as discussed above. For some others, it could be a difficult process for many reasons.

To begin with, different carriers have different interests in air cargo and different mindsets, partly because cargo revenue is a relatively small percentage of the total revenue for some carriers. Second, if some ground functions are outsourced, it is difficult to control the quality of the service provided to just one airline let alone multiple airlines. Third, cargo configurations among airlines are much different than passenger configurations. Fourth, the customer base can be very different for different carriers. Some are retailers and work with individual shippers; others are wholesalers and work with freight forwarders. Some transport low profit-margin cargo just to fill space and increase market share, others segment the marketplace and select the customers and products based on their profit margins. Fifth, there are significant differences among products, costs, philosophies, and computer infrastructure.

While many of these obstacles have existed on the passenger side, carriers have been able to resolve them because of the higher priority given to the transportation of passengers than to the transportation of cargo. Therefore, an important question is that while a strategic alliance among passenger carriers for the transportation of cargo can offer shippers more destinations and alternative routings in case of problems, can the alliance integrate its marketing, operations, and investments in technology to lower operating costs for alliance partners and raise the quality of delivery for shippers?

For airlines willing to integrate their networks, sales forces, price-service offers, handling process, and information technology systems, cargo alliances provide extraordinary opportunities.[15] Integration of network provides mutual benefit for partners as well as for shippers. For partners, it reduces duplication of flights that, in turn, trims operating costs and increases utilization. For shippers, integrated networks provide greater access to world markets as well as more services and or more comprehensive departure and arrival times. Combining regional sales forces provides an incentive for the integrated sales force to sell the combined capacity of the alliance rather than creating competition among partners. The process saves costs of having multiple sales forces within the same region. Integrated price-service offers are critical for the movement of cargo in origin-destination markets that require transportation by two or more partners. Obviously, a given product must be available from all partners at the same price and with the same level of service. Similarly, handling processes (pick-up, delivery, tracking, tracing, and so forth) need to be consistent among the partners. And, perhaps most important, the existence of common information technology systems will not only facilitate the standardization of handling processes but also the harmonization of total demand for the services of and total capacity provided by the alliance partners.

Air cargo services are provided by a broad spectrum of businesses. Historically, scheduled passenger airlines have transported cargo in the bellies of their passenger aircraft as well as some airlines that have also provided freighter services. Traditionally, most, but not all, of these carriers have viewed air cargo as a commodity business and sold the service on the basis of price resulting in a decline in yield. They provided airport-to-airport services and have sold their capacity on the wholesale market to intermediaries—the freight forwarders. Examples of the three largest carriers in this category are Lufthansa, Korean Air, and Singapore Airlines. Charter carriers also provided similar airport-to-airport services but focused more on the needs of shippers who had special consignments. At the retail level, freight forwarders sell transportation door-to-door to individual shippers. Then, there are agents that sell ancillary services such as customs clearance and final delivery services at the wholesale or retail levels.

Beginning in the late 1970s, a new type of air cargo service providers entered the marketplace—the integrators or express carriers.

Integrators provide value-added services (door-to-door service, time-definite products, and performance guarantees).[16] Until recently, the integrators focused on the overnight envelope traffic and the small package business. Four of the largest carriers in this category are FedEx, UPS, DHL, and TNT.

Figure 6.2 provides one view of the business models of these groups of service providers. The integrated carriers focused on the high value to weight consignments that required minimum transit time—often overnight service. Scheduled airlines and freight forwarders transported medium value to weight shipments taking longer times than integrators to make the delivery. Charter airlines were generally speaking at the upper right-hand corner with respect to the shipment value to weight ratio and transit time axes. Figure 6.3 shows another way to portray business models of these service providers—the relationship between shipment weight and the scope of services proved. Examples of the scope of services provided include airport-to-airport at one end, door-to-door service (sometimes even desk-to-desk service) in the middle, to totally managing the supply chain of the shipper. Figure 6.3 also includes another category of service providers—postal services. While there has been overlap in the services

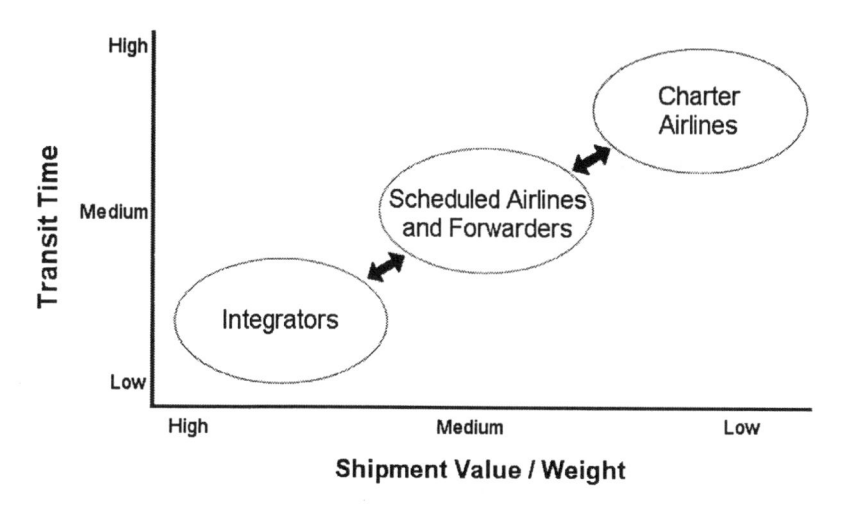

Figure 6.2 Air cargo service providers and one interpretation of their business models

provided by different groups of business entities, in recent years boundaries have become blurred between airlines, freight forwarders, and express operators. Some integrators are moving in on the territory of some scheduled airlines and vice versa. Similarly some scheduled airlines are moving in on the territory of freight forwarders and visa versa. And in some cases all three—integrators, scheduled airlines, and freight forwarders—are raising the scope of services provided by moving up the supply chain management ladder.

The involvement of postal services can partially be explained by need of the postal services within Europe to prepare for postal deregulation. Fearing the potential loss of their monopolies in the

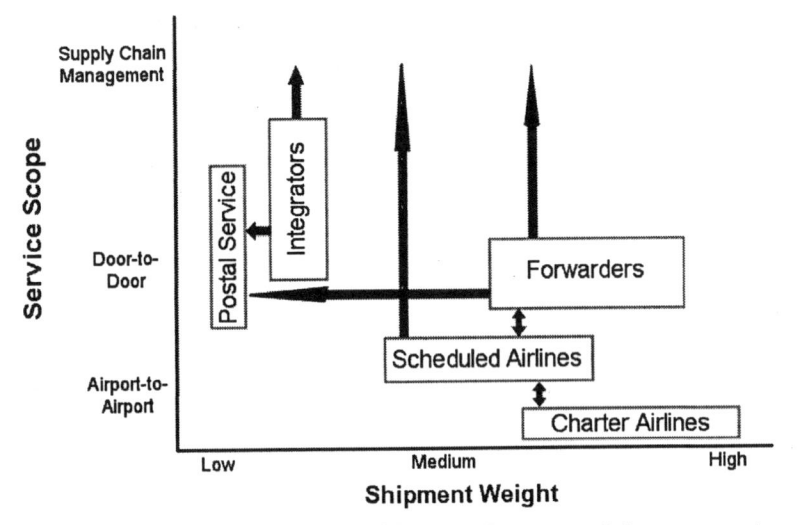

Figure 6.3 Air cargo service providers and a second interpretation of their business models

transportation of letters, some postal services began to acquire interests in freight forwarders and integrated carriers to become pan-European parcel carriers. Examples include the financial interest of Deutsche Post—Europe's largest Post Office—in the global integrated carrier DHL and multinational freight forwarders Danzas and AEI. Deutsche Post also has an alliance with an information technology business and Lufthansa (that, in turn, also owns part of DHL) making it a true giant logistics business. This conglomerate alliance is ideal for meeting the customers' emerging needs of finding innovative solutions to logistics through fast, dependable, comprehensive, flexible, and worldwide service. The conglomerate can provide such services by coordinating and integrating products, handling and sales processes, and IT systems. Similarly, but to a much smaller extent, the Dutch Postal service acquired an interest in the integrated carrier TNT.[17]

One explanation for the migration of integrated carriers toward the transportation of packages and heavier freight is the slow but continuous decline in the rate of growth in overnight envelope traffic due the availability of fast and cheap mode of electronic transmission, a little due to fax machines and more due to e-mail and e-mail attachments. FedEx's average shipment size has been reported to have increased from six pounds to 7.8 pounds in the past five years.[18] Therefore, even though the envelope business is not decreasing—only its rate of growth is declining—integrated carriers are moving toward the transportation of small packages. The second reason for focusing on the transportation of small packages is the high growth rate of this segment. Some shippers who previously might have tendered their small shipments to freight forwarder to be consolidated into large shipments to receive a break in price are now tendering the same small shipments to integrated carriers at higher rates in return for faster delivery times. It cannot, however, be assumed that the integrated carriers have lost interest in envelope traffic. It is partly to maintain interest in that segment and partly to try to capture a larger share of the small package business that some integrated carriers have formed partnerships with postal authorities worldwide. The recent agreement between the US Postal Service and FedEx is just one example in North America.

In order to make even further inroads in the small-package business, some integrated carriers have begun to add more trucks in their networks to expand their less-than-truckload services. There are two reasons for this strategy. First, traditional truckers have been increasing their activities in intra-regional markets in North America and Europe, regions with extensive and very good road network. In the US, for example, truckers have begun to provide a broader spectrum of expedited services. Second, trucks are ideal for moving shipments cost effectively in markets under 800 miles through their one, two, and three-day products.

The decision of some scheduled carriers to migrate a little toward the integrated carrier business model results partly on the need to improve yield. Figure 6.1 showed the decline in passenger yield resulting partly from a steady improvement in technology and productivity and partly from an increase in competition.[19] In recent years, passenger carriers have been focusing more and more on the potential contribution of cargo to offset partially the higher level of decline in passenger yields. The additional focus on the contribution of cargo, however, encouraged in turn the belly-service providers to reduce cargo rates even further. In order to get out of this spiral, some scheduled carriers began to examine the higher-yield segment dominated by the integrators. And it is part of this segment of the market that the traditional carriers have begun to capture by introducing time-definite products. Lufthansa Cargo was, for example, the first major carrier to convert its entire product portfolio to time-definite services.[20]

It is too early to tell the final outcome of this migration process. For example, integrated carriers, who have traditionally focused on very reliable transportation of documents and small shipments, have been moving toward the transportation of heavier shipments transported by scheduled airlines. However, this transition may not prove to be easy as their current mode of operation is based on handling relatively standard products (for example, small shipments with bar codes) and with relatively standard operations (with respect to transit times and the use of hub-and-spoke systems). Although their handling process is labor intensive, their yields are also higher. Handling heavy freight will prove to be different. To accept heavier shipments would require integrators to increase efficiency to offset the lower yield. It could also be difficult to compete with the coordinated strategies of some scheduled airlines and freight forwarders. Examples of such coordinated strategies include the use of the Internet to plan logistics, and time-definite products with lower transit end-to-end time—a real opportunity for heavier shipments.

The non-integrated service providers (both forwarders and scheduled airlines) operate across a wide spectrum of situations—handling, for example, very small shipments as well as very large shipments. They, too, are trying to move in the business of integrated carriers. This migration is a little easier because forwarders, for example, can use the services of multiple airlines not to mention other modes of transportation such as trucks. Consequently, they have the advantage of providing customized services to shippers and consignees.[21] However, the non-integrated carriers could also have significant difficulty in matching the information technology capability of integrators because of legacy systems and in some cases legacy mindsets of non-integrated carriers. For example, airlines within an alliance can communicate well on the passenger side for making reservations on each other's flight. Yet, such a capability is not widespread within the cargo segment.

Bridling Technology to Redefine the Air Cargo Market

Aircraft

Airlines began to use jet aircraft for transporting freight in the mid-1960s, employing either new or converted versions of passenger configurations of the Boeing 707 and 727 as well as the Douglas DC-8 and DC-9. Most of these aircraft were operated by both cargo and passenger airlines to transport between airports cargo tendered mostly by freight forwarders. Wide-body aircraft—Boeing 747s and the DC-10s—began to enter the marketplace in the late 1970s. In the 1980s, integrated carriers entered the marketplace with the Boeing 727s and the Douglas DC-8s. Large international passenger carriers concentrated on the use of wide-body aircraft such as the Boeing 747s. From the late 1980s other aircraft manufacturers began to offer a broad spectrum of freighters with varying capacities and ranges exemplified by the small British Aerospace BAe-146s to the large Airbus A300s to meet the growing and varying needs of cargo service providers. For example, not only was there a significant increase in global trade but even more important was the enormous interest in just-in-time manufacturing processes. However, cargo service providers had historically been interested in acquiring converted versions of passenger aircraft to obtain lower costs to compensate for their lower utilizations, especially of the integrated carriers.

In recent years, due partly to the high growth within the air cargo industry—particularly within integrated carriers—partly because of superior economics of the medium-capacity wide-body aircraft, partly because of their higher volumes, and partly because of their dispatch reliability, cargo carriers have begun to order new aircraft such as the Boeing 767-300 and the Airbus A300-600. One forecast says more than 3000 freighters will be delivered between the year 2000 and 2019. Approximately two-thirds will be for regional operations and about one-third for long-haul flights. Within the regional operations category, about one half will be with a capacity of less than 30 tons and one half with a capacity between 30 and 50 tons. In the long-haul category, about 40 percent of the freighters are forecast to have a capacity between 50 and 80 tons and about 60 percent a capacity more than 80 tons. Finally, within both categories of regional operations about 90 percent of the aircraft are expected to be converted from passenger configurations and only about 10 percent are expected to be purpose-built freighters. Within the long-haul categories, almost one-half of the aircraft are expected to be converted from passenger configurations and the other half to be new designs.[22]

The integrators—such as DHL, FedEx, TNT and UPS—decided to replace their lower-volume narrow-body freighters such as the Boeing 727s with the intermediate capacity wide-body freighters such as the Airbus 300s and the Boeing 767s. This appears to be consistent with their decision to penetrate the small package business. First, it is more efficient to accommodate small packages in wide-body freighters. Wide-body aircraft are more efficient even for the transportation of documents and mail that tend to have higher density than the small packages.[23] Acquisition of wide-body equipment opened up more capacity for integrators to compete even more with freight forwarders. Rather than compete directly with the forwarders for the heavy freight, integrators encouraged targeted shippers of consolidated freight to break their consignments into smaller-size shipments and tender them to the integrators. They have been developing attractive price-service options to achieve their objectives.[24] This segment of the traffic is large. For example, one estimate is that only 15 percent of the international air freight is heavy and cannot be divided into smaller packages.[25] The strategy of integrators to migrate in this direction makes sense.

The demand for freighters will continue to grow despite the fact that (a) the size of the air freight market remains relatively small compared to the passenger market, (b) about one half of the freight still moves in the bellies of passenger aircraft, and (c) excess belly capacity has encouraged marginal cost pricing. First, growth in the air cargo business has been and is expected to continue to outpace growth in the passenger business. Second, even though shippers can receive lower rates for belly space, the quality of transportation varies significantly when transported in the bellies due to higher priority given to passenger considerations. Third, prior to the 11 September incidents, passenger load factor had been increasing, placing a constraint on the amount of cargo that could be accommodated in the bellies after the extra luggage had been accommodated due to the higher load factor.

Capacity of future freighters will also encompass a much larger range—from conversions of small narrow-body aircraft such as the Boeing 737s/Airbus 320s to newer versions of the Boeing 747-400s and the Airbus 380-800Fs. At the low end, manufacturers could even bring to the market twin-engine propeller-powered freighters to replace the Boeing 727s and at the high end the payload capacity of the A380-800F could approach 350,000 pounds.[26] At each capacity (and range) level, carriers will be able to select the freighter that offers the optimal combination of payload and operating costs. Take, for example, the case of the A380-800F at the top end. A carrier like FedEx could replace two MD11s with one A380 in some trans-Atlantic markets and four with one in some trans-Pacific markets such as Memphis-Hong Kong.[27] The A380 freighter would not only provide a lower unit cost payload-range capability, but also the potential for creating—primarily in Asia—new hubs and expanding existing hubs to provide nonstop service between North American, European, and Asian hubs. "wet-lease" operators such as Atlas could operate the A380 for a consortium of airlines. The other advantage of new freighters is that would comply with the environmental requirements. A significant part of the freighter movement is at night, making additional curfew restrictions at such airports as Frankfurt extremely detrimental. It would not be cost effective to operate freighters from near-by airports based on the fact that about 40 percent of the cargo transported by a carrier such as Lufthansa Cargo is carried in the bellies of passenger aircraft.[28]

Information Management

Emerging technology represents both a threat and an opportunity for the air cargo industry. For example, the Internet could force out those freight forwarders who do not add value. Not only are some cargo airlines beginning to offer time-definite products directly to shippers (particularly those shippers who are willing to bring their cargo to the airports) but also it is possible that the prices paid to airlines by freight forwarders for capacity may now become more transparent as a result of the Internet as well as through the development of online forwarders.

Consider, a somewhat different threat. Small cargo players cannot justify the investment in cutting-edge information technology, which is needed to avoid potential breakdowns in communications and coordination due to the involvement of multiple parties involved in the transport chain. Yet, such information technology is exactly what is needed for customer satisfaction.

A third example of a threat for traditional scheduled airlines and freight forwarders comes from integrators with state-of-the-art information technology capability to provide real-time logistic solutions (for example, significant compression of cycle times with guaranteed delivery times and real-time tracking and tracing). Legacy systems of traditional airlines and freight forwarders are not compatible with the Internet technology.

The fourth area of threat is the Internet's effect on prices. Availability of information on prices can and has created competition among various categories of service providers. The Internet facilitates not only greater transparency in the information on rates but it also enables customers to go into a bidding war more easily, activities that will put an even greater pressure on yield. Further declines in yield will require service providers to reduce their costs even more, hopefully through the implementation of emerging technology.

Emerging technology, therefore, provides both challenges and opportunities for air cargo industry. Let us consider the Internet and e-commerce on the opportunity side. The Internet provides information that is global, economical, real time, and transferable between a buyer and a seller (data and video). The Internet provides benefit to the shipper, the consignee, and the airline. Everyone can communicate via the Internet. For shippers and consignees, the Internet provides online accessibility and a cost-effective information exchange system among various parties in the transportation chain (worldwide) to enable management of different shipments at different service levels. For carriers, the Internet can help to reduce costs and enhance revenue. Let's look at the cost side. It costs about two dollars for UPS to track a package if the request comes in via a telephone call. It only costs 10 cents if the tracking is done online.[29] FedEx has the most widely used website. It is reported that almost three-fourths of its shipments (more than one million per day) are processed online without paperwork. Costs can also be reduced by reducing ground time (for example, activities relating to customs clearance and the generation and processing of air waybills). And a reduction in the number of places a shipment stops will reduce the costs of warehousing the shipment.

E-commerce is more than a website. E-commerce can basically be described as a multi-channeling capability in which a shipper can do business with the airline in a way that is best for the shipper. For example, a shipper can participate in online exchanges, participate through his or her own website, or work through the website of the service provider. It can add value in the supply chain management through the widespread availability of rapid communications and the capability to connect dispersed databases. Consequently, e-commerce can reduce transaction costs, increase market penetration, and improve competitive position.

Emerging technology is also playing a significant role in the consolidation/alliance process described above. First, electronic links can make the alliances really effective through the development common websites. The IATA Cargo 2000 initiative can provide value through the standardization of information among the different service providers. Technology can help in the customs clearance area, for example, by pre-clearing shipments while they are still en route. Technology can improve some ground handling processes. While there is no problem to track a shipment at the consolidated level, it is a problem to track it as an individual shipment. While air waybills contain information on the weights of shipments, this information is sometimes incorrect. The incorrect weight problem could easily be unintentional, for example, due to the rounding process. Consequently, it is possible that airlines may not be collecting the correct amount of money for the shipments being transported. Moreover, incorrect weight can be a real problem with respect to aircraft weight and balance calculations.

One novel concept proposed in the area of emerging technology is the implementation of customer relationship management at the industry level.[30] Transportation of cargo is different. Its success depends on cooperation and coordination among numerous diverse and sometimes fragmented business entities between the shippers and consignees. These entities include various types of airlines, forwarders, airports, truckers, and customs. The enabling technology—such as wireless communications and smart cards—is available to implement CRM at the cross-industry level. The issue, however, is not the lack of available technology but rather the integration of existing technology using existing interfaces. The second obstacle is the lack of willingness to share data. The third potential obstacle may be the requirement to move away from the stand-alone mainframe technologies toward the outsourced Web-based technologies.

Figure 6.1 showed a continuous decline in the air cargo yield. This decline has been the result of (a) competition from the marginally-costed belly capacity, (b) more productive aircraft with lower unit operating costs, (c) the proportionately high volumes of cargo moving in the Asia-Pacific markets where the longer distances lead to lower yields, and (d) price-negotiating power of freight forwarders who control large volumes of shipments. The key question now is how service providers can reduce costs in line with the reduction in yield and at the same time not only maintain service but also enhance service to meet the growing expectations of shippers and consignees. Simple solutions have not proven to be effective. For example, some cargo providers outsourced their ground handling activities, with the selection of the company based on the lower price per pound handled. Such cost reduction strategies can be ineffective if the company being outsourced to be is then forced to keep its costs low by hiring lower-paid workers and or by reducing the level of service provided.

In the long run, the future for air cargo can be bright. In the increasingly competitive environment, many combination airlines may find that revenue from belly cargo may contribute more and more to their overall income. For the integrated carriers, globalization is leading to an increasing demand for air cargo in general and door-to-door overnight services in particular. And it is the judicious acquisition and implementation of emerging technology (such as the Internet and e-commerce) as well as established technologies (such as revenue management) that can assist both types of carriers to maintain the rates, reduce costs, and maintain service levels through economic cycles.

Notes

[1] Boeing Commercial Airplanes Group, World Air Cargo Forecast, 2000/2001, September 2000, p.1.
[2] Nelms, Douglas W, "Perspective reality," *Air Transport World*, September 2000, p.97.
[3] Taverna, Michael A., "Lufthansa Spearheads Alliances In Freight, Express and Logistics,"*Aviation Week & Space Technology* 27 August 2001, p.59.
[4] Nelms, Douglas W, "Perspective reality,"*Air Transport World*, September 2000, p.98.
[5] Ibid. p.98.
[6] Convey, Peter, "A battle for e-cargo,"*Airline Business*, September 2000, p.58.
[7] Convey, Peter, "Cargo on-line," *Airlines Business*, February 2000, p.78.
[8] Nelms, Douglas W., "Perspective reality,"*Air Transport World*, September 2000, p.98.

Forces Transforming the Air Cargo Market 151

[9] "The New Global Cargo concept,"*Aviation Strategy*, June 2001, p.11.
[10] *Airline Business*, March 2001, p.77.
[11] Ott, James, "New Pressures Set Off Alarms for Air Cargo," *Aviation Week & Space Technology*, 27 August 2001, p.48.
[12] Boeing Commercial Airplanes Group, World Air Cargo Forecast, 2000/2001, September 2000, p.2.
[13] Boeing Commercial Airplanes Group, World Air Cargo Forecast, 2000/2001, September 2000, p.2.
[14] Sindemann, Holger, "Cargo alliances—a way out of the crisis?" *Aviation Strategy*, November 2001, p.16.
[15] Sindemann, Holger, "Cargo alliances—a way out of the crisis?" *Aviation Strategy*, November 2001, p.15-16.
[16] Convey, Peter, "Cargo on-line," *Airlines Business*, February 2000, p.79.
[17] Convey, Peter, "A battle for e-cargo,"*Airline Business*, September 2000, p.58.
[18] Page, Paul, "Express Boxed In," Air Cargo World Online, April 2001, p.1.
[19] Boeing Commercial Airplanes Group, World Air Cargo Forecast, 2000/2001, September 2000, p.1.
[20] "The New Global Cargo concept,"*Aviation Strategy*, June 2001, pp.9-10.
[21] Clancy, Brian and David Hoppin, "The 2000 MergeGlobal Air Cargo World Forecast," Air Cargo World Online, May 2000.
[22] Airbus Industrie, Global Market Forecast 2000-2019, July 2000.
[23] Page, Paul, "Express Boxed In," Air Cargo World Online April 2001, p.5.
[24] Page, Paul, "Express Boxed In," Air Cargo World Online April 2001, p.6.
[25] Page, Paul, "Express Boxed In," Air Cargo World Online April 2001, p.6.
[26] Smith, Bruce A., "Boeing Explores Cargo Possibilities for Sonic Cruiser," *Aviation Week & Space Technology*, 27 August 2001, p.54, and Sparaco, Pierre, "Cargo Trends Spur Airbus To Pursue More A380F Orders,"*Aviation Week & Space Technology*, 27 August 2001, p.55.
[27] Saparaco, Pierre, "Cargo Trends Spur Airbus To Pursue More A380F Orders,"*Aviation Week & Space Technology*, 27 August 2001, p.58.
[28] "The New Global Cargo concept,"*Aviation Strategy*, June 2001, p.11.
[29] Jindel, Satish, " E-Commerce: Misconceptions Go Online," Air Cargo World Online February 2001, p.7.
[30] Mills, Colin, "re-inventing the air cargo industry," Presentation made at the ACI Cargo Conference in Sharjah, 19 March 2001.

Chapter 7

Business Structures and Processes to Capitalize on Emerging Technology

Some statements just aren't compelling news. These include: an airline needs to (a) react quickly to the changing marketplace, (b) provide value to its customers, or (c) promote continuous innovation to deal with the changing marketplace. It is also not compelling to hear that airlines need to quickly find ways to deal with the converging forces of liberalization, privatization, globalization, and increasingly dynamic and uncertain marketplace. However, what *is* news is that technology now enables an airline to develop management philosophies and business structures and systems that can respond efficiently and effectively to these and other changes. Most airlines are already sophisticated users of such technology as yield management, network management, and recovery systems. However, they can now capitalize on different areas of technology to improve their business structure and systems.

This chapter starts with a highlight of a potential vision and role of technology and discusses such key questions as: (1) is technology an enabler or driver of business strategy?; (2) should airlines deploy basic technologies or exotic technologies?; (3) what are some of the key concerns relating to the assessment of investments in technology? The second section highlights some potential contributions of technology with respect to business philosophies, structures, and systems. Examples include value-based planning, the role of the chief information officer, legacy systems, integrated data systems, and knowledge management.

Vision and Role of Technology

Most strategies being adopted by traditional airlines deal with more or less the same attributes—hub-and-spoke systems, strategic alliances, high-margin passengers, lower distribution costs, and so forth. One factor that differentiates winners is the existence of conducive organizational structures and practices that enable an airline to execute its strategy effectively. Problems, therefore, are likely to be the result of flawed executions, not flawed strategies. Poor execution of strategies can be the result of inadequate systems for measuring the performance of strategies, inadequate organizational structures for executing knowledge-based strategies in an increasingly uncertain business environment, inadequate measures of future financial performance—measures that relate to value-creating strategies.[1] Consider, for example, the manner in which numerous decisions are made daily by many people at many levels. Each person's decision may appear to be quite reasonable from that person's perspective. However, because these individuals have different objectives, information, tools, and incentives, their actions can be irrational since the consequences of all these individuals are not likely to be aligned with the company-wide objectives of the airline. Moreover, flawed execution of strategies can also be the result of how different decision makers view the role of technology.

Is Technology an Enabler or Driver of Business Strategy?

Should business strategy drive technology strategy or should technology strategy drive business strategy? In theory, business strategy should drive technology strategy. In practice, however, it also may be possible for technology strategy to drive business strategy. Thus, it may be more beneficial to view technology as both an enabler and a driver of business strategy. For example: An airline can start with a business strategy and then develop an appropriate technology strategy. However, the airline should also be ready to modify its business strategy to derive the full benefit from the available technology. Here are some examples.

Let's start with a straightforward business strategy to reduce distribution costs. For a traditional large airline the distribution costs through an airline's own website can be as low as one-fourth the costs of using travel agents—the traditional distribution channel. The enabling technology would presumably then be a Website and the Internet and the business strategy would be for the airline to direct as many transactions to its own website as possible. However, in computing the costs and benefits of this strategy the airline would also need to modify its business strategy by developing such activities as cross-channel interactions and a commission structure that supports a multi-channel strategy. Consider a passenger who may go to the Web to get information on a trip. This use of the Web, however, may merely eliminate the need for certain information from the call center or the travel agent. The passenger may in fact still contact the call center to get additional information to complete the transaction, especially if the itinerary is complex. Therefore, a passenger using the Web should be able to contact the call center while still on the Web and have an airline's representative begin communications over the Web. Such a process would then optimize the value of the cross-channel interaction. Consequently, the optimal use of a particular channel may require more than technology. Finally, after visiting the Web and, possibly even after contacting the call center, a passenger may still end up going through a travel agent. Consequently, the relationship between business strategy and technology strategy is iterative, and one that requires total integration among all channels.

Another business strategy could be improving customer service. An airline must first decide the approach for improving customer service (high touch, high-tech, or a combination) and the areas of customer service (pre-flight, in-flight, or post-flight). Suppose that an improvement is desired in all three areas and management selects the use of the high-tech approach. The enabling technology could then be a combination of the Internet, mobile communications, speech recognition, biometrics, and self-service machines. However, the successful implementation of the available technology requires a careful alignment between business strategy and technology strategy. Following are a few examples.

1. The website needs to be functional and user friendly. Functionality means that the design should take into consideration a passenger's total requirements and provide comparative information. User friendly means that it should be easy for a passenger to make flight inquiries and bookings, select and change seats, and print boarding

passes. However, the technology strategy relating to the Website needs to be aligned with the business strategy. For example, the technology strategy would depend on whether the business strategy is to use the Website simply to provide information, or to provide transactional capability, or provide the capability to provide personalized service.
2. Customer service can be improved through the deployment of speech recognition technology during difficult operational conditions, because it is convenient and user friendly for the passenger and because it is scalable for the airline. Scalability means that the system can adjust to accommodate the change in the volume of activity without any deterioration in the response time for the customer. Again, the application of technology must be considered in conjunction with business aspects such as language and cultural considerations. Certain words and certain tones portray different meaning in different languages and different cultures. In addition, the language and culture aspects, in turn, affect the brand.
3. The use of biometrics technology can improve customer service by improving the customer processing activities at airports. However, its use raises the question of acceptance by passengers and government authorities. In addition, it requires coordination among multiple entities such as airlines, airports, and government authorities. (See the discussion on Simplifying Passenger Travel in Chapter 2.) Finally, one must recognize that airlines and airports operate with different objectives. The airline may wish to have passengers in and out of the airport as quickly as possible but the airport may wish to keep passengers at the airport as long as possible given the income that is derived from airport shops, restaurants, and other services.
4. Self-service kiosks at airports can improve customer service for a selected segment of customers. However, the CUSS (Common Use Self Service) capability is required to gain full benefit from their use. Given the limited space available at the large congested airports, this requirement means that all airlines should be able to use a common set of self-service machines. Although technology is available to fulfill such a requirement (based on the experience of bank ATMs), its use requires airlines to adopt common standards and to collaborate to create a global infrastructure within which everyone

can compete.[2] Such collaboration then brings into focus other issues such as security, privacy, and brands. Finally, airlines also need to conduct rigorous cost-benefit analyses, since the implementation of these machines does not eliminate the need for traditional methods for processing passengers. It is simply an alternative method. Many alternative methods raise the question, should airlines provide them in light of the pressures to reduce costs and improve security?

As the preceding examples show, the issue is not so much as to whether technology is an enabler or a driver of business strategy—it is both—the issue is the need to align technology strategy with business strategy. Consider, for example, the deployment of two business strategies at the same time: (a) attract more high-fare passengers; and (b) use the Internet to sell distressed inventory at very low fares to raise load factor. The Internet has already proven to be an effective channel for attracting low-fare passengers. However, the use of this channel depresses passenger yields. Thus, if the business strategy is both to go after both high-fare passengers and low-fare passengers, which would be better: to use the Internet to attract low-fare passengers or to establish a low-fare subsidiary to attract low-fare passengers? The two decisions have different ramifications with respect to staff and processes for handling passengers.

Next, suppose the primary business strategy is to identify and retain high-value customers. Now the enabling technology might include sophisticated segmentation techniques, CRM, and profitability analyses (discussed in Chapters 3 and 2, respectively). An airline would need to use these technologies to meet the changing priorities and expectations of its customers, from identification through acquisition to the retention of high-value customers. However, as pointed out in Chapter 3, implementation of CRM initiatives requires major changes in business thinking, practices, processes, and resources. However, even some of the airlines who decided to implement CRM are not fully cognizant of the complete set of requirements for CRM processes. The project cannot be implemented piecemeal. A strategy must exist to establish an enterprise database. There must be training for employees. There must be strategy to embrace technology to connect the airline's legacy systems and its contemporary systems that are likely to be based on the Internet Protocol (IP). In addition, there must be significant technology to connect all the partners in a strategic alliance to improve a passenger's total origin-to-destination experience. Consequently, if the business strategy is to attract and retain high-value flyers, the airline must not only move from the static website to applying many of the e-business initiatives (discussed in Chapter 4), but also to align its business strategy with its technology strategy.

In conclusion, then, technology is both an enabler and a driver of business strategy; for example, the Internet enabled the reverse price auction sales implemented by priceline.com. But it was the guaranteed overnight delivery business model that led Federal Express to develop the tracing/tracking technology. Theoretically it may be more meaningful to say that an airline should first develop a visionary business model and then obtain the necessary technology to execute its strategy (as in the case of Federal Express). But from a practical point of view this may not be timely and cost-effective in light of the availability of so many emerging technologies such as mobile communications that provide opportunities to proactively manage customers. Both models for the deployment of technology can be valid as an enabler and as a driver for delivering value in the sense that, for example, while it is doubtful that any single company could have asked technologists to develop the Internet to implement its business strategy, a company can develop a business strategy that builds on the Internet.

Basic Technologies or Exotic Technologies?

Emerging technology has a different meaning for different people. Some people look at emerging technology with respect to its application to improve the basic core functions of an airline. Others look at it as a way to provide exotic services. Consider, for example, just one emerging technology (mobile communications) and just one application (the passenger check-in process). One airline might use this technology to provide its roaming customer service agents at an airport with hand-held devices that can be used to check-in passengers when the check-in lines get long. A second airline might decide to be at the leading edge of this technology and to use it to check-in passengers from their cars as they approach the airport. A third airline may want to be at the bleeding edge of this technology and use it to not only check-in the passenger as he or she drives up to the airport but also allow the passenger to change seats, be upgraded to first class using a frequent flyer mileage award, receive a bar-coded signal to act as a boarding pass, receive information about the assigned gate, and be informed of the location of a specific parking space open in the parking lot nearest the departure gate of the passenger.

Deregulation in some parts of the world and liberalization in other parts have made the airline industry highly competitive and resulted in lower profit margins. Consequently, there is strong pressure to increase the profit margin by increasing the revenue, decreasing the costs, or both. Prior to the 11 September attack, most of the large airlines appeared to be moving toward the use of exotic technologies to make travel easier and to increase revenue through the use, for example, of CRM and e-business projects. A few airlines remained more interested in deploying the available basic technologies to achieve a significant improvement in their financial performance, for example, within the area of revenue accounting. The 11 September incident increased interest in the value of these basic technologies, at least in the near term. Consider just two areas of revenue accounting, fare audit and proration.

In a given market and in the economy class of travel, there exists a broad spectrum of published fares. The full fare (with no travel restrictions) can be as high as five or six times the lowest discount fare (with many restrictions such as minimum duration of stay and advance purchase requirements). It is possible that an intermediary may sell a lower fare when a passenger has not met the necessary conditions associated with that lower fare. The situation may occur unintentionally or intentionally. Either way, such situations cause airlines to lose significant amounts of revenue. One group of consultants estimates that the airline industry could be losing up to 3 percent of its sales revenue due to the use of erroneous fares.[3]

Because of the prohibitive cost of verifying each ticket, airlines have typically audited only a small percent (3 to 5 percent) of their total tickets, although they may verify a much larger percent of the tickets sold by a particular intermediary or in a particular market. Until recently, the audit process has been manual, therefore less reliable, and very costly. Technology is now available (based on the automated logic) to verify 100 percent of the tickets quickly and cost effectively. Such technology can not only identify the erroneous tickets, but also, the source of sale, the type of violation and the amount of revenue involved. Technology can then automatically produce a debit memo in the appropriate amount to be sent to the source of the violation.

Proration is another form of ticket audit. When a passenger travels on two or more airlines using a single ticket issued by one airline, a complex set of rules divides and shares the revenue in different currencies among the airlines providing the service. Proration is a very complex activity because it involves the application and interpretation of numerous rules and conditions, some of which are established at the industry level and some that are unique to individual airlines providing the service. Consequently, the activity is not only time consuming but also subject to errors. Again, technology is now available to provide timely and accurate billing involving a broad spectrum of tickets and fares.

Intelligent Assessment of Technology

Unlike in such assets as aircraft, it is almost impossible to assess investments in technology in any definitive manner. The difficulty arises with respect to all of the traditional three elements involving the technology initiative: costs, benefits, and the investment. The assessment problem makes it very difficult to develop a compelling business case for many of the emerging technology initiatives.

Some costs are straightforward such as the costs of computers, wireless application protocol (WAP), radio frequency identification (RFID) devices, and self-service kiosks. However, other costs are much more difficult to assess. These include the costs of changing the processes. Consider, for example, the deployment of self-service kiosks for simple tasks similar to bank ATMs such as those for checking-in passengers with simple itineraries, seat upgrades, and seat changes. Although use of these machines can free staff time for complex tasks, it is difficult to compute accurately the reduction in the number of staff required because of the deployment of kiosk technology. In the case of biometrics technology, its deployment at airports involves high installation costs of hardware. Since there are multiple beneficiaries, how should the costs be shared among the airlines, airports, and government agencies?

Similar problems exist on the computation of benefits. Wireless technology is invaluable for providing information to time-sensitive frequent travelers. How can such benefits be quantified? Similarly, it would be difficult to compute the total benefits of in-flight connectivity that enables passengers to send and receive e-mail, surf the Web, and watch live TV. While it may be relatively easy to compute the revenue produced by these services (from user charges), it is almost impossible to compute the benefits from the use of such technology. Just as benefits cannot be measured from the viewpoint of passengers who use them, they also cannot be measured easily from the viewpoint of an airline which might consider the availability of these services to be one of the differentiators among the choice of airlines. Consider next the benefits of technology that helps an airline to improve its interface with employees and customers. Integrated data systems are a case in point. They are essential ingredients for the implementation of CRM initiatives in which the customer touch points can now be used by airlines to perform security and verification tasks and record the information in the integrated data system. In fact, such information may even need to be recorded in an industry-wide data system. Therefore, unlike costs, benefits of such data systems not straightforward. Similarly, how does one compute the benefits of upgrading legacy systems to make them compatible with Internet Protocol? Finally, how does one compute the benefits of technology that can provide an increase in bandwidth and an improvement in connectivity among systems that can lead, in turn, to the use of u-commerce (discussed in Chapter 4)?

Finally, it is not just the costs and benefits that are difficult to compute for the deployment of emerging technology, it is just as difficult to estimate the required investment. Following are just two potential explanations. First, experience shows that many systems become outdated quickly, requiring upgrades and replacements. Therefore, airlines have always underestimated the total investments required to make and keep many systems fully functional. Second, some systems have limited applications when they are first introduced. Take the case of the technology available to provide wireless access to the Internet and e-mail in airport lounges. At the present time, the amount of wireless traffic within an airport environment limits the extent of this service. Moreover, since different passengers use different service providers for their Internet services, it would be very difficult for an airline to provide access to all wireless users.

Airlines have been typically allocating about three percent of their total budget to technology. Despite the proclaimed benefits of emerging technology, they have been apprehensive about increasing their costs. The situation has become worse after the 11 September incidents. On the other hand, the need to invest in technology is growing.

1. In the future, corporations are likely to track the performance of airlines with respect to their claims. If, for example, the delays incurred by a particular airline are constantly above the industry level, that airline could lose the contract with the corporation monitoring the performance. Alternatively, a corporation could track performance and expect compensation if performance fails to match expectations and promises.
2. The need to improve connectivity with customers, employees, and strategic partners is growing. It is the breadth and depth of this connectivity that can enable airlines to innovate in a number of marketing and operational areas such as new channels of distribution, self-processing capability at airports, e-commerce, and seamless travel involving multiple partners. In light of the increased concerns for safety, the need for connectivity must now include other parties such as airports and numerous government agencies.
3. Technology can play a very important role in improving the security aspect of the airline travel. Examples include the use of biometrics technology to screen passengers more comprehensively and the use of data warehousing and mining capabilities to develop effective customer profiles. Technology can also keep track of where passengers and their bags are at any given time and where they have been.
4. Some airlines have considered the potential for advertisements on their websites as a way to generate additional revenue. Advertisements can provide an additional source of revenue as long as such initiatives do not make the airline become unfocused. Similarly, technology can also enable airlines to generate more money from cross selling (hotels, cars, and so forth) activities. Again, cross-selling initiatives only make sense if they add value.
5. Technology can lead to innovation[4] which can lead to new business models. Chapter 2 described IATA's initiatives with respect to SPT. Chapter 3 discussed the value of CRM initiatives. Chapter 4

described e-business initiatives to improve the total experience of a passenger. Some business writers have identified three specific components of customer satisfaction: performance, price, and personalization.[5] Technology can facilitate the identification of innovative ways to achieve improvements in all three areas.

Assessing investments in emerging technologies has always been a difficult task for the airline industry. It may even be more difficult for managements of airlines owned by governments, which may investigate investments in technology from a different perspective. The motivation tends to be different. These owners tend to take fewer risks and want to be more certain of expected returns in terms of value as well as time horizon. In assessing investments in technology, it might also be useful to keep in mind that cycle times are coming down while competition increases, and customer patience shortens. Moreover, technology itself can facilitate the decision-making process by minimizing risk when both time and analyses are limited.

Business Philosophies, Structures, and Systems

It is a generally accepted business concept that the key attribute of a customer-centric organization is customer satisfaction, driven by customer value. As stated in the previous section, some business writers have concluded that there are three key components of customer value: performance, price, and personalization. It therefore stands to reason that in order to develop a customer centric organization, it is necessary to adopt a business philosophy, organizational structure and systems that enable management to provide customer value by developing products and services that customers will evaluate with respect to performance, price, and personalization.

Value-Based Planning

Some strategy consultants describe today's airline planning structure as complex, lacking transparency, and built on artificial organization lines.[6]

Complexity can easily be observed from the existence of layers upon layers of control. Such an organization structure is neither conducive for entrepreneurship—and therefore innovation—nor for controlling costs. The lack of transparency is evident from the fact current organizational structures make it difficult to allocate both revenue and costs to individual products. For example, flights operating to and from the top tourist destinations may post losses if they are used by passengers who redeem their frequent flyer awards on these routes. However, some of the revenue earned on flights operating to and from business destinations used by frequent flyers ought to be allocated to flights to and from tourist destinations where frequent flyers redeem their rewards.

The existence of artificial organizational lines is evident from the fact that the responsibility for customer relationships is split among multiple organizations. For example, a passenger forms expectations through the touch points involving the sales and marketing department whereas the fulfillment of these expectations is in the control of the operations department. Chapter 2 discussed the need for cross-functionality that is even more critical in the implementation of value-based planning. See Figure 7.1. Moreover, artificial organizational lines are tied to budgets and budgets are tied to history and fair share, rather than what it would take to get the strategy implemented. Figure 7.1, that highlights the value of cross-functionality at three levels (operational, tactical, and strategic), can provide value in decisions relating to budgets.

One group of consultants recommends that airline management should consider developing value-based strategies and structures by reducing the complexity in their decision-making process, by developing transparent accounting systems, and by promoting entrepreneurship environment both within and among the different groups.[7] Take, for example, the case of entrepreneurship spirit within a group, say the passenger check-in unit. Such a spirit would lead this group to keep a scorecard that shows trade-offs among such attributes as on-time performance, operating costs, and customer satisfaction. The basic idea behind value-based planning is the identification of units (reflecting different touch points) where value is generated and identification of value drivers within individual units or at individual touch points. Customer satisfaction and brand loyalty are two examples of value drivers.

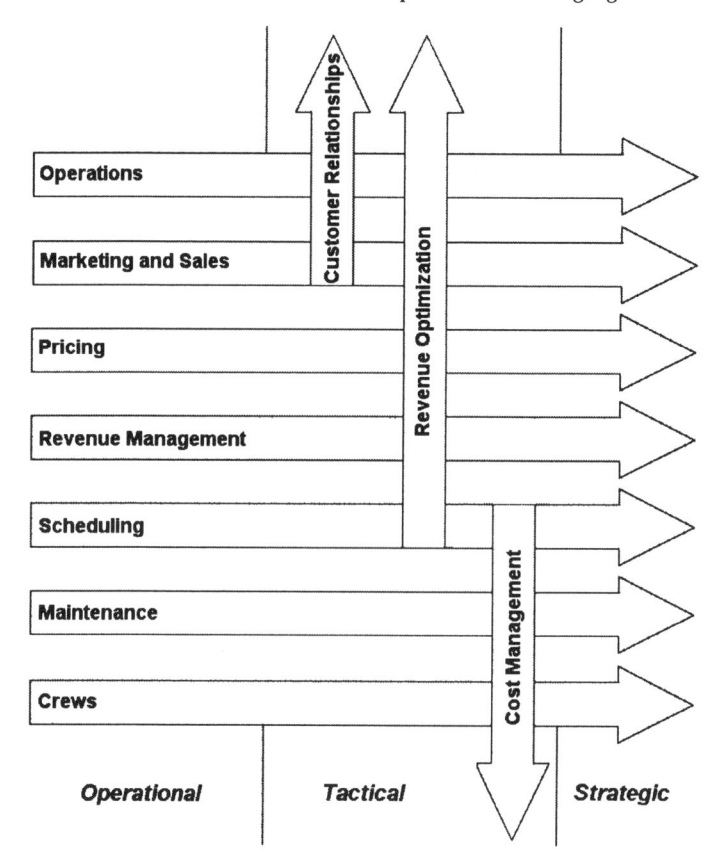

Business Structures and Processes to Capitalize on Emerging Technology 165

Figure 7.1 Cross-functional integration

Within the airline industry strategic alliances have emerged as a major force that can enable individual airlines to compete in global markets. Among alliances, centralized network planning, centralized sales forces in selected countries, and integrated IT systems are three examples of units where value is created.

1. The value of a central network-planning unit can easily be observed in a case where two airline partners with very different operating

cost structures operate flights in a given market. Clearly, it would add value if all flights were to be offered in the market under consideration by the airline with the lower operating costs. However, such a scenario can only materialize if there is a transparency, trust, and an entrepreneurship spirit.
2. As above, the success of a centralized sales force requires the implementation of a transparent budget, a transparent set of rules of the game, and recognition of the unit generating the value and the unit receiving the value. Consequently, there is a need to identify operational drivers at each touch point—the ease with which a passenger can get information at a website and the relevance of the information, the amount of time spent in a check-in line, and so forth.
3. An integrated IT system is needed if partners are to provide integrated services across all members (reservations, check-in, e-tickets, frequent flyer programs, and so forth). Take the case of frequent-flier programs. It is not just the common rules for earning and redeeming miles that are important to passengers but it is also important for passengers to receive appropriate recognition and privileges across all members. Even within one airline, let alone within an alliance, overlap of responsibilities can cause problems. Therefore, it is important to centralize the necessary data and information, employee empowerment/incentives, and measurement metrics. Unfortunately, the metrics relating to customer service are not very precise or structured. In addition, the difficulty with employee empowerment is that there are cultural issues. Employees with some airlines based in Asia tend to work according to strict rules, with little flexibility. The rules have in fact become even more complex. For example, an agent may now have on screen an extensive menu. This menu may indicate to the agent, for example, to offer the passenger this if the passenger delayed, or that if the bags are lost. The existence of these rules eliminates flexibility that is needed by the agent to evaluate the situation and make an appropriate decision on the spot.

The principles of value-based planning can be applied to all functions. Take the case of distribution. The first generation of global distribution systems focused on putting information into the hands of professional intermediaries, for example, travel agents. The second generation focused on adding value by putting information directly into the hands of consumers. However, this changed environment does not imply the end of the agency business, a sector that will always provide value through personal relationships with certain segments. Consequently, it is not the travel agency business itself that will end, but rather it is the percentage of business handled by as well as the compensation schemes of travel agents that will change.

The value-based planning also applies to decisions involving outsourcing. An airline should always outsource activities that are commodities and do not add value—for example, a communications network—but not activities that touch customers. Under this scenario, vendors of technology need to show that they not only understanding the airline's core business and processes but also that they can add value by sharing the risk of adopting new technology and innovation In addition, technology providers should not over promise and under deliver, a strategy that has affected negatively the airline industry's expectations.

The current focus in business strategies is to concentrate on core business (fly passengers safely and on time) and reduction of costs. Both areas of emphasis imply a greater amount of outsourcing or cooperative development of technology, resulting in more opportunities for Application Service Providers (ASP) relating to such activities as recruitment, mobile services (information on flights such as the delays cancellations and gates), data processing and analysis (warehousing and mining), revenue accounting, and fare management (auditing and proration). Outsourcing and cooperative development could be extremely valuable at this time. Andy Hayward of South African Airways explains the situation. Senior management is not likely to accept a proposal that involves significant investment in technology without a cast iron business case, particularly during times when an airline is simply trying to survive. However, without ongoing investments in technology, an airline's competitiveness and operational effectiveness could decline, the costs of its legacy systems could increase, and emerging technology initiatives such as CRM could be postponed. Even during normal times, few airlines individually can afford to fund the ambitious projects when (a) there is the risk of 'reinventing the wheel,' (b) there is a strong possibility that the initial competitive edge may not exist in the long run, and (c) rapid advances in technology can easily make today's state-of-the-art solutions obsolete.[8]

For the reasons mentioned above, selective outsourcing and cooperative development can provide significant savings in costs and eliminate a number of the problems associated with the integration of legacy systems. However, there are some concerns relating to the use of ASPs that need to be addressed: (1) handing over competitive and valuable data to an external company; (2) potential lack of consistency; (3) reliability; and (4) inability of the external company to provide customized service, for example, corporate business rules applied to data warehousing and data mining. The first three concerns can be resolved fairly easily. The fourth concern is a little more difficult to resolve. Customization requires flexibility and that raises costs. A resolution of these concerns can, however, provide significant benefits with respects to lower costs of development and maintenance and the availability of latest technology.

Chief Information Officer

Probably the most important critical success factor in the effective deployment of technology is the selection and the operating environment of a chief information officer (CIO). The right individual working in an appropriate organizational setting can help senior executives understand the role of technology and enable an airline to (a) drive its business strategy, and (b) facilitate change. CIOs have multiple conflicting demands placed on them. The chief executive officer (CEO), for example, may want to improve financial results with technology that is good, fast, and cheap. Unfortunately, it is difficult to acquire technology with all three attributes. It is only realistic to obtain any two of the three attributes. Functional executives want technology that makes each of their own jobs easier. Customers want technology to make the products more convenient. For them, CIOs want technology that is fast, flexible, scalable. It is not reasonable to expect technology to meet such divergent needs of all potential users. In addition, when these users do not get their unrealistic expectations met, they question the value of technology. Many of these dilemmas can be resolved by selecting a CIO with the right background and creating an appropriate working environment for the CIO.

Historically, CIOs have generally been looked upon as experts in technology—executives responsible for meeting the needs of airlines in the areas of data processing and communications. More recently, CIOs have also been made responsible for the development and management of functional databases and contracts with technology vendors. A CIO must, however, be a business strategist as well as a technology expert. One writer refers to the qualifications of a CIO as "technology-proficient but business-savvy."[9] Such a background will enable an airline to link its business strategy with its technology strategy and provide a balance between the two strategies. Moreover, since an airline CIO also needs to be a catalyst for change, the CIO with experience on the business side is more likely to be able to facilitate change not only by demystifying technology but also by showing (a) realistic benefits and costs, and (b) innovative ways to get real value out of technology. It is a given that the CIO needs to know technology to understand today's extremely complex technological environment. However, it is the business experience of the CIO that will enable him or her to understand the business strategy and how to leverage it through technology. Furthermore, a CIO with business experience can show how technology (databases, metrics, tools, techniques, and so forth) can be used to increase value for customers, employees, and shareholders.

Many senior executives do not understand how technology works and do not have confidence in technology in light of the fact that many technology projects have historically been behind schedule and over budget. Such disbeliefs in the value of technology may be the result of one or more of the following factors:

1. The CIO may have been given a different set of standards at the beginning of the technology project from the set of evaluation standards employed at the end of the project.
2. While some technology initiatives can produce benefits in the short term, many technology projects produce benefits in the long term. Examples include the CRM initiative discussed in Chapter 3 and the middleware systems and integrated databases discussed later in this chapter.

3. Airlines have almost always considered technology as a cost item rather knowledge enabler and driver of change through the management and dissemination of revealing and actionable knowledge—knowledge that can be used to develop and execute a meaningful corporate strategy.
4. In most cases, it is very likely that management did not conduct realistic cost-benefit analyses of technology. Rarely do managements include the full costs of change, not only those involving the costs of new employees and training but, more important, costs involved in changing processes, attitudes, and behavior of employees.

In addition to selecting a CIO with the correct background, it is also necessary that he or she be placed in an organizational structure so that the CIO can drive the maximum advantage from technology. From this perspective, the first requirement is that CIOs must have their own functional units and not be part of other functional units such as marketing, finance, or administration. This requirement helps a business to own major technology projects and view and manage them like all other business projects. Moreover, this requirement can enable the CIO to make a decision on the operational aspects of the information technology function to be outsourced, freeing up the CIO to focus on strategy and the value-added components of the function. The second requirement is that the CIO must be part of the top management team, so he or she can contribute strategically. This requirement implies that the CIO must not only report directly to the CEO but that he or she be elected to the airline's board. Such an operating environment will not only produce a productive functional unit but a functional unit that facilitates the integration of other functional units. These points are well articulated by one writer in a statement, "the 'I' in CIO stands not only for information but also, if properly supported and profiled, for intelligence, integration, and innovation."[10]

Legacy Systems

An important decision that an airline CIO must make is related to the replacement or modification of legacy systems—the original mainframe and other customized computer systems used by airlines that embody old ways of doing things. In the 1970's, most airlines used two different technology platform systems—developed by IBM and Unisys—to support their reservation systems and the SITA network to provide access to the systems by the travel agencies. Both systems continue to be reliable, accurate, and cheap to operate.[11] However, these systems cannot accept easily the contemporary applications that take advantage of the Internet revolution. Moreover, the legacy systems are not relational data based. Later, airlines began to implement networks for internal functions such as payroll, personnel, finance, and operations. As each functional system tended to be customized for a particular functional area and often developed using proprietary infrastructures, airlines ended up with parallel and unconnected systems and networks. Consequently, the resulting systems (both reservations and functional) had two basic problems. First, they could not share the information among themselves. Second, they were not compatible with the Internet Protocol.

Consider, for example, Mr. Smith showed up at the check-in counter and wanted to be upgraded. The agent sent a message to the central reservation system, which denied the request. However, until now, there was no way for the central reservation system to receive information from the integrated data system, saying that this passenger had been inconvenienced three times (lost baggage, missed connection due to a late arriving flight, and denied accommodation in business class due to an oversold situation), and that he was at the threshold of abandoning the airline. If that information were available to the agent or to the reservation system, perhaps the answer to the upgrade request would have been positive.

Now, technology known as middleware has become available, providing the connectivity between the old and new data formats. Middleware can link an agent's screen (or call center) to the central reservation system and other disparate systems relevant for the particular case. This would capitalize on vast quantities of data available across worldwide networks. Many airlines continue to be constrained by their legacy systems even though technology has become available to either acquire totally new systems (with technology that incorporates open and scalable architecture),[12] or upgrade the existing systems with middleware technology. Part of the reluctance is based on the costs of totally new systems as well as the costs of rejuvenating the legacy systems. A second part of the hesitation can be laid on the fear of the unknown relating to the new systems when the old systems are working reasonably well. Consequently, the established airlines have been trying to move forward by making incremental changes to their legacy systems. Unfortunately, the incumbents need to change at a faster speed than they are comfortable with because some of their competitors have made the move for higher speed.

Integrated Databases

In the airline industry, data has traditionally been stored in dozens of unconnected databases that existed in various functional units such as marketing, sales, finance, and operations. Moreover, almost all of these unconnected databases existed at the headquarters of each airline. People in the field kept their own ad hoc pieces of data and had very limited access to the data kept by their own functional divisions at the headquarters. There, each functional unit used its own database to make recommendations to executive committees responsible for making strategic decisions. The executive committees (composed of such groups as chairpersons, presidents, executive vice presidents, members of the board, and so forth), faced great difficulty in making decisions, often due to the inconsistency of the data used by different units. Moreover, the data collected provided little value to the people in the field who needed to make tactical decisions virtually in real time.

Airlines have always felt the need to consolidate their disparate databases into a comprehensive system so that everyone could have a single, unified view of the data. Such an objective has remained a dream for two basic reasons. First, technology was not readily available to handle the enormous quantities of data generated and received from external sources. Second and perhaps even more important, functional units wanted to control their own data. Since information was power, most functional units did not want to share the information and, in turn, the power. In recent years, two major developments have increased the need to consolidate databases and make the information available to all the staff, both at the headquarters as well as in the field worldwide. First, airlines now have even more data to deal with as a result of the Internet (and the e-commerce) revolution. Second, increase in competition has led to low profit margins and, in turn, the need to make efficient and effective tactical decisions in the field, worldwide and in almost real-time basis.

Fortunately, in recent years technology can enable airlines to have data warehouses (consolidated multidimensional databases) that meet the aforementioned requirements. Figure 7.2 provides one example of a corporate data warehouse that shows a dozen examples of data sources—ranging from marketing and sales to operations. Going a little deeper, obviously, flight coupons and air waybills, as well as bookings and reservations would be the fundamental building blocks of the database or data warehouse. The Marketing Information Data Tapes (MIDT) is another example of a critical source of data. This source contains huge amounts of information on reservations made through a certain number of Computer Reservation Systems (CRS).[13] A data warehouse can also contain data mining capabilities—techniques to analyze data to identify trends and patterns. These are the integrated data systems. Integrated means that data collected is at the enterprise level (the entire airline). System means that it has two components, a data warehousing capability and a data mining capability. An integrated data system can assist management in a number of areas such as the six shown in Figure 7.2. It can enable an airline to analyze and explain past events, forecast the outcome of current decisions, and, consequently, provide management with some control over a future event. Consider the following example.

An integrated data system can help an airline to analyze and explain the erosion in its passenger yield—for example, due to defection of high-margin passengers because of poor performance relating to both aircraft and baggage. Such a system can also identify thresholds for different categories of passengers. For example, the system can provide information that a certain segment of frequent flyers defected after three delays involving missed connections or flight cancellations. Based on this information, the airline can now identify passengers likely to defect with the next operational problem and the kind of discounts or special offers that might preempt their defection. The system can score the information to predict passenger behavior and make such information available to any relevant user, worldwide and at any time.

174 *Driving Airline Business Strategies through Emerging Technology*

Sources of Data
- Sales
- Websites
- Flight Experience
- Customer Service
- Revenue Management
- Finance and Accounting
- Frequent Flyer Programs
- Human Resources
- Alliance Partners
- Baggage Claim
- Operations
- CRS

Integrated Database

Uses of Data
- Asset Management
- Customer Care Management
- Marketing and Sales Campaign Management
- Customer/Product Profitability Analysis
- Product Development and Evaluation
- Employee Care Management

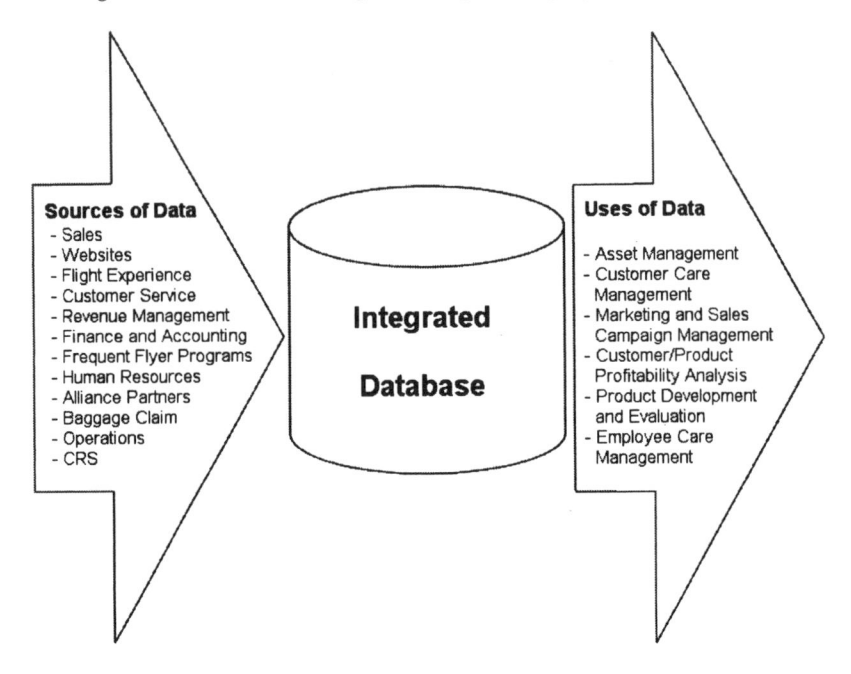

Figure 7.2 An integrated data system

From an organizational point of view, the development and management of the integrated data system needs to be the responsibility of the CIO. From the business perspective, the CIO will ensure that there is a three-fold purpose of the integrated data system.

1. Such a system needs to provide vital insights into the business such as the identification of the most important customers and their needs, status of the financial and operational performance, and benchmarked comparison of products. These insights must be based on a single, unified view of information using organizationally consistent data and a consistent set of business rules. Such a system can now also play an important role in matching the security-related information obtained at various passenger touch points with the information stored in the integrated data system—for example, an analysis of booking and ticketing behavior.
2. The system must support the tactical and strategic decision-making processes by (a) facilitating employees to understand customers,

operations, and partners and (b) empowering employees and providing them with a timely single, unified view of information. For example, it was mentioned earlier that customer satisfaction has three components: performance, price, and personalization. An integrated data system can help employees in the field understand the third component and its relationship to the other two components. For instance, some passengers want personal treatment and recognition. However, they also want the airline to respect their privacy. Some passengers want the airline to not just satisfy their needs but also anticipate their needs. An integrated data system can help employees understand customers. However, to satisfy customers, it is also necessary for management to empower employees.
3. Users must be able to use the integrated data system in both the proactive and reactive mode. In the proactive mode, the system helps to make decisions that affect future operations. In the reactive mode, the system helps to explain the events that have already taken place. The data collected from a website (one of the sources shown in Figure 7.2) can be reactive or proactive. For example, information can be gathered about the usage of websites, not only automatically, but, in fact, through simple questionnaires that can be set up to seek additional information from these sites.

The old databases focused on the needs of senior executives at headquarters who used the data to make strategic decisions. Now we need data marts for individual business areas containing real-time information, not just data, to make tactical decisions at the field level as well as strategic decisions at the executive management level. Consider the situation when an aircraft is about to leave the gate and flights with connecting passengers on board have not landed. How should the local manager make the decision? The local manager needs detailed information on passengers who will miss this connection. What is the value of these passengers? The system must also provide a value of passengers who may defect if they are made to wait any longer. The system must provide quickly and accurately the values of both groups of passengers and the level of service received by both groups.

From the technical perspective, first the system needs to be flexible enough to accept new data, remove obsolete data, and update existing data. Second, the system must make the customer data readily available to any user at any time. Third, the system must be able to answer a broad spectrum of queries in an equally broad spectrum of formats. In other words, users should be able to ask questions in their own functional context and receive an answer that reflects all cross-functional considerations. See Figure 7.1. Fourth, the system needs to be scalable in that performance must not deteriorate when there is an increase in the volume of data, workload, or number of users. The system should be able to handle tens of thousands of users any day and any time of the day. Consequently, a large international airline could easily require the system to handle double-digit terabytes of data and support tens of thousands of concurrent users. One terabyte is one trillion bytes of data—the number of characters in two million books.[14] Finally, it should be possible to load the data into the warehouse directly from a broad spectrum of sources such as those shown in Figure 7.2.

The value of an integrated data system cannot be underestimated. Today's environment stresses such attributes as decentralized decision-making processes and collaborative planning. The strategic and tactical use of integrated databases can help an airline leverage its existing information to gain a competitive advantage through improved business control (at the strategic level) and customer service (at the operational level). For example, the use of such a system at the strategic level can help an airline to identify the business model that is optimal for it, namely, which segments, what product features, what channels of distribution, and what price. Customer files contain information on customer value and the service to be provided. Therefore, at the tactical level, this information can be helpful in analyzing incoming calls to reservation centers to determine the optimal routing of the call and treatment of the passengers making the call, based on the entire transaction history of the caller as a guide to the customer service agent. This information can improve customer service, reduce costs, and enhance revenue.

Knowledge Management

Airline executives are fully aware of the evolutionary thought leadership regarding management philosophies. Here are some examples.

1. Fifteen or twenty years ago, Total Quality Management was in vogue. This philosophy attempted to relate the purpose of work to customer satisfaction by using feedback to improve business processes. It also assumed that employees at all levels would take on the responsibility of finding new ways to do their jobs by defining problems and finding solutions. One of the key drivers of the TQM philosophy was organization learning that, in turn, depended on the existence of effective group or team communications.[15] The TQM philosophy was originally promoted in the manufacturing industry and Hewlett-Packard is reported to be one of the first American companies to adopt the technique.[16]
2. About 10 years ago, it was the "Balanced Scorecard." This philosophy claimed that businesses that relied strictly on the use of financial measures in a management system caused organizations to do the wrong thing, for example, by promoting short-term behavior at the expense of long-term value. This conclusion was based on the belief that financial measures are lag indicators in that they report on outcomes resulting from actions taken in the past. The "Balanced Scorecard" concept did not discard the use of financial measures; rather, it supplemented them with measures that reflected leading indicators—measures of future financial performance that, in turn, meant measures of strategy. Examples of drivers of future financial performance cited were measurements related to customers, internal business processes, and growth perspectives.[17] A number of leading firms such as Mobile Oil and Chemical (Chase) Retail Bank are reported to have adopted the use of the "Balanced Scorecard" concept.[18]
3. Recently, some business writers have been promoting the "Six Sigma" concept as a method to improve the effectiveness and efficiency of businesses by improving customer satisfaction through focus and management of the organization's processes. The "Six Sigma" concept itself refers to a statistically derived performance target to achieve only 3.4 defects for every million activities or opportunities.[19] A number of prominent corporations such as

General Electric and Allied Signal are reported to have adopted and benefited from the "Six Sigma" concept.[20]

4. Also recently, some writers have been examining the feasibility of using knowledge-based strategies that depend on the organization's intangible assets such as customer relations, informational databases and skilled and knowledgeable employees.[21] The feasibility of these knowledge-based strategies depends on the degree to which an organization believes in such concepts as change management and knowledge management and the degree to which the firm has acquired the skills, tools, and culture to implement such concepts.

Despite their popularity in business books, the airline industry in general does not appear to have adopted any of these management philosophies. Following are some potential reasons for the lack of interest by the airline industry.

1. The airline industry is in the service business and its operations are very complex, exemplified by the existence of dozens of customer touch points, a widely spread geographic network, a round-the-clock operation, and a highly unionized workforce working in an extremely complex web of rules.
2. Customer satisfaction is not an easy task in an industry whose operations are affected by many factors that are outside of the control of the industry. Examples include weather and capacity of the aviation infrastructure (airports and air traffic control) and the increase in airport processing time required for additional security.
3. A number of the concepts mentioned above rely on technology that had not matured at the time when the concept was in vogue.
4. Some managements find the knowledge management concept to be abstract, that is, difficult to define and difficult to measure. Therefore, it is difficult to compute a ROI on projects that involve knowledge management.

While it would seem reasonable to assume that competing in the emerging environment will require airlines to focus on process measurements and knowledge management, most airlines continue to be organized in the traditional centralized, functional, budget, and slow-reacting structure. This situation is, however, likely to change due to the impact of powerful forces such as the emergence of strong low-cost and low-fare airlines and now the 11 September incidents. In addition, those airlines seeking to achieve a source of sustainable competitive advantage in non-traditional areas could find one in the development and management of knowledge.

Knowledge management is essentially about: (1) converting data into information and information into internal knowledge about the marketplace;[22] (2) applying the knowledge to understand market trends; (3) converting knowledge into action.[23] Although it is difficult to distinguish clearly among data, information, and knowledge, here is one attempt to show an example of the conversion process.

Depending on the airline, between 5 and 15 percent of the passengers traveling in economy class are forced to buy the highest-priced seats due to their inability to plan ahead. These fares can be four to six times higher than the lowest discounted fares and do not carry any restrictions such as advance purchase and minimum length of stay. The discounted fares are heavily restricted. However, traditionally there was virtually no difference in the service received by the passenger paying the higher fare during the pre-flight, in-flight, or post-flight phases. The difference between the full-economy fare and the heavily discounted fare was justified on the basis of flexibility of travel associated with the normal full-economy fare.

Data can only show the number of passengers who paid the full economy class fare. At some airlines, airport staff and flight attendants converted this data into information from the comments received from these passengers that they were very dissatisfied with the lack of differentiation in the service received at airports and during the flight by passengers paying full fares. This information was then converted into knowledge that a significant portion of these passengers were finding a number of loopholes to get around these somewhat artificial restrictions to obtain lower fares. Knowledge was converted into action when some airlines began to offer a separate part of the cabin with additional services for these passengers. Other airlines took a different approach. Some increased the legroom for a few rows and allocated these seats to the full-fare passengers. Others began to assign the more desirable seats to the higher-paying passengers.

Two types of knowledge are needed in an organization, knowledge to improve the performance of a business when the organization knows its objectives and knowledge to help the organization develop new objectives and the associated strategies.[24] It is the pursuit of both types of knowledge that provides a business its competitive advantage. To take advantage of knowledge, however, it is first necessary to create knowledge—knowledge that is necessary to make and keep the organization customer- and business-centered, that is, having a goal of meeting customer and shareholder expectations. Knowledge can come from the external environment such as customers and or the internal environment such as employees. Once knowledge has been acquired and stored, it can be distributed and applied in a collaborative framework to make business decisions.

There are well-defined steps for implementing knowledge management.[25] The first element of knowledge management is the availability of a good technology infrastructure—an integrated data system that is readily accessible, broad-based electronic chat rooms, intranets (internal corporate Internet), and extranets (intranets that include other companies such as strategic alliance partners). Next, there is a need to have knowledge management strategy encompassing such elements as

(a) the approach for converting data into action
(b) the level of granularity of knowledge
(c) key attributes for finding relevant knowledge (indexing and retrieval systems)
(d) incentives for sharing knowledge and disincentives for hoarding it
(e) the need for cultural change that promotes such concepts as organization learning, employee empowerment, and an open environment where employees know and see how they contribute to the goals of the business.

Finally, while airlines are experienced in managing tangible assets (such as aircraft), now they must learn to manage intangible assets such as those mentioned above—customer relations, informational databases, and skilled and knowledgeable employees. In addition, it is the skillful management of these intangible assets that can lead to a non-traditional source of competitive advantage.

Organization learning, mentioned above, is a key component of knowledge management. The need for organization learning has come about because businesses are becoming more complex and more global (in the case of airlines, as a result of strategic alliances). In this context, knowledge has become more fragmented and uneven, transactions require urgency, real-time information needs to be widely distributed to employees working within the 24/7 framework, and cycle times are compressed.

E-learning, in turn, is a key component of organization learning. Some people associate computer-based training with e-learning. E-learning is much more than computer-based training; it is about learning to solve problems by processing information online. The Internet, intranets and mobile communications can enable the information to become available to all employees in almost real time. The integrated data system can ensure that the employees can access the information in any format—information that is accurate, organized, and consistent. From passengers' viewpoint this information is now very critical as they expect the airline to deliver a safe and secure trip rather than just a hassle-free trip. From an airline's viewpoint, information management and the associated technology can leverage employees' skills to areas that add the most value for the airline.

There is a compelling rationale for e-learning. In many airlines, learning activities are restricted mainly to the technical areas such as flying and maintaining the aircraft. In addition, even here it is training, not learning. Learning is much broader than training. It includes, for example, generating and disseminating information (that reflects the collective wisdom of the company) to support the performance of a business.[26] Training, on the other hand, is the delivery of instructions (associated with a skill or competency) while learning is the delivery of information (associated with knowledge management).[27] There is a need for both online training to improve the skill or competency of an individual employee and online learning to improve business practices and capabilities. Airlines need to do both more online training and even more online learning. However, as one business writer puts it, it is not just new technology to learn with, it is a new approach to learning.[28] For example, e-learning can also be used by customers in gaining more information about the products that they are purchasing.

Just as there is a need to align business strategy with technology strategy, there is also a need to align the e-learning strategy (creating and sharing knowledge) and the business strategy because e-learning involves not just the acquisition of technology but also changes in a number of other areas such as culture and justification of costs.[29] A change is required in the corporate culture not only for the business to provide to the right people the right information at the right time but also for the employees to create, apply, and share information. The value of improved communications has become even more evident after the 11 September incidents. Moreover, not only do employees need to have easy access to relevant, accurate, and usable information, management needs to establish new metrics if learning and performance are to be tied to business results. United Parcel Service, for example, defined four point-of-arrival metrics[30] (customer satisfaction, employee relations, competitive position, and time-in-transit). Such metrics can be used to assess how different units of an airline create value for current and future customers. Justification of costs means that the business must develop a comprehensive business case justifying the higher costs of e-learning (higher than the traditional costs associated with class-room training) to the higher benefits from an enhanced business performance.

As mentioned above, technology has played a vital role in enabling airlines to manage their tangible assets such as aircraft. Technology can now play an equally important role in enabling airlines to manage their intangible assets such as customer relations, informational databases, and skilled employees. It is the effective management of such intangible assets that can lead to a non-traditional competitive advantage. However, management of intangible assets requires more than the acquisition of technology; it also requires airlines to change their organizational structure, processes and culture to let technology not only be an enabler of change but a driver of change. Full potential benefit of technology cannot be derived if airline managements continue to have the traditional organizational structures, budgeting processes, fragmented databases, and dysfunctional positions for their chief information officers. Full exploitation of emerging technology also requires organizational structures that enable the deployment of a collaborative approach in the planning and execution of strategy—an approach that itself can be facilitated by technology.[31]

Notes

[1] Kaplan Robert S. and David P. Norton, *The Strategy-Focused Organization: How Balanced Scorecard Companies Thrive in the New Business Environment* (Boston, MA: Harvard Business School Press, 2001), Chapter 1.
[2] Schapp, Stephen and Richard D. Cornelius, "U-Commerce: Leading the New World of Payments," A While Paper by Visa International and Accenture, San Francisco, July 2001, p.7.
[3] Telephone interview with Ashish Malhotra, Kale Consultants, New York, 3 September 2001.
[4] Shapiro, Stephen M, *24/7 Innovation: A blueprint for surviving in an age of change* (New York: McGraw-Hill, 2002), pp.129-30.
[5] Mittal, Banwari and Jagdish N. Sheth, *Value Space: Winning the battle for Market Leadership*, (New York: McGraw-Hill, 2001), p.11.
[6] Telephone interview with Nikolas Hermann, Roland Berger Strategy Consultants, New York, 8 September 2001.
[7] Ibid.
[8] Based on a dialogue with Andy Hayward, Chief Information Officer, South African Airways. January 2002.
[9] Bell, Michael, "Empowering IT leaders," *Airline Business*, October 2001, p.96.
[10] Ibid.
[11] Kirby, Alex, "Taking responsibility," *airline info tech*, Summer 2001, pp.9-12.
[12] Hearns, Paul, "Bridging the technology gap," *airline info tech*, July-August 2000, pp. 41-3.
[13] Clark, Paul, *Buying the Big Jets* (Aldershot, Hants, England: Ashgate Publishing, 2001), p.43.
[14] Klebnikov, Paul, "The Resurrection of NCR," *Forbes*, 9 July 2001, p.72.
[15] Kochan, Thomas A. and Michael Useem, *Transforming Organizations* (Oxford, England: Oxford University Press, 1992), p.378-9.
[16] Fahey, Liam and Robert M. Randall, *The Portable MBA in Strategy*, (New York: John Wiley & Sons, 1994), p.126.
[17] Kaplan Robert S. and David P. Norton, *The Balanced Scorecard: Translating Strategy into Action*, (Boston, MA: Harvard Business School Press, 1996), p.18.
[18] Kaplan Robert S. and David P. Norton, *The Strategy-Focused Organization: How Balanced Scorecard Companies Thrive in the New Business Environment* (Boston, MA: Harvard Business School Press, 2001), pp.3-7.
[19] Harry, Mikel and Richard Schroeder, *Six Sigma: The Breakthrough Management Strategy Revolutionizing the World's Top Corporations*, (New York: Doubleday, 2000).
[20] Ibid.
[21] Kaplan, Robert S. and David P. Norton, *The Strategy-Focused Organization: How Balanced Scorecard Companies Thrive in the New Business Environment* (Boston, MA: Harvard Business School Press, 2001), p.2.

[22] Boyett, Joseph H. and Jimmie T. Boyett, *The Guru Guide to the Knowledge Economy: The Best Ideas for Operating Profitably in a Hyper-Competitive World* (New York: John Wiley & Sons, Inc., 2001), p.97.

[23] Pfeffer, Jeffrey and Robert I. Sutton, *The Knowing-Doing Gap: How Smart Companies Turn Knowledge into Action*, (Boston, MA: Harvard Business School Press, 1999), pp.248-51.

[24] Hatten, Kenneth J. and Stephen R. Rosenthal, *Reaching for the Knowledge Edge: How the Knowing Corporation Seeks, Shares & Uses Knowledge for Strategic Advantage,* (New York: American Management Association, 2001), p.2.

[25] Tiwana, Amrit, *The Knowledge management Toolkit: Practical Techniques for Building a Knowledge Management System,* (Upper Saddle River, NJ: Prentice Hall, 2000).

[26] Rosenberg, Marc J, *e-Learning: Building Successful Online Learning in Your Organization,* (New York: McGraw-Hill, 2001), p.11.

[27] Ibid., p.29.

[28] Ibid., p.31.

[29] Ibid., pp.32-3.

[30] Kaplan, Robert S. and David P. Norton, *The Strategy-Focused Organization: How Balanced Scorecard Companies Thrive in the New Business Environment* (Boston, MA: Harvard Business School Press, 2001), p.21 and Chapter 9.

[31] Methner, Bruce E. and Christopher J. Rospenda, "Airline Strategy in a Digital Age: What Does "e" Mean to Me?," Butler, Gail F. and Keller, Martin R. (eds), *Handbook of Airline Strategy,* (New York: McGraw-Hill, 2001), p.403.

Chapter 8

A Call for Action

The landscape of the airline industry has been changing for some time. From a long-term perspective, even though passenger traffic has grown at a rate much higher than the growth in economies, the growth in the industry's real revenue has been less than the growth in economies. See Figure 8.1.

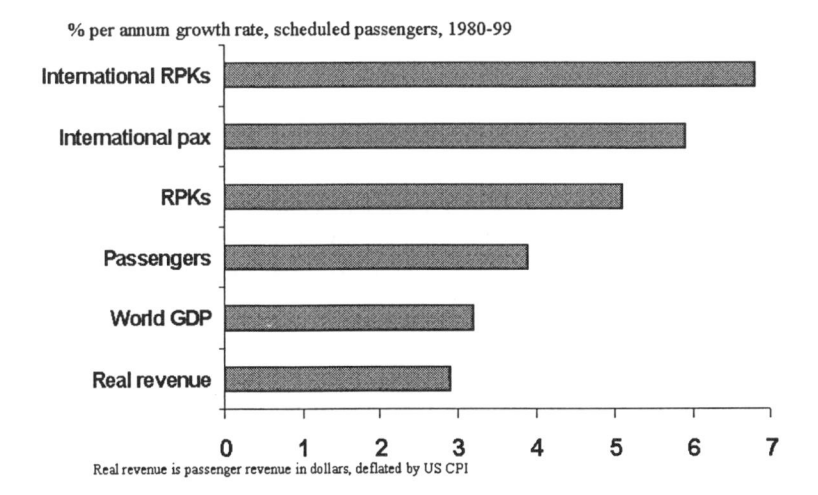

% per annum growth rate, scheduled passengers, 1980-99

- International RPKs: ~6.7
- International pax: ~5.8
- RPKs: ~5.1
- Passengers: ~3.9
- World GDP: ~3.2
- Real revenue: ~2.9

Real revenue is passenger revenue in dollars, deflated by US CPI

Figure 8.1 While passenger traffic growth has outperformed GDP growth, real revenue growth has not
Source: British Airways

Even worse, since World War II the net profit margin of the industry has been close to zero. From a mid-term perspective, the industry has been experiencing increasing amounts of segmentation—low-fare airlines at one end of the spectrum, second-tier flag-carriers and independent carriers in

185

the middle, and the mega-carriers and their strategic alliance partners at the other end of the spectrum. From an even shorter-term perspective, the industry has experienced a significant decline in its revenue as a result of (a) slowdown in the economies, (b) defection of some high-yield passengers revolting against the huge differentials between the full fares paid by on-demand business travelers and deep discount fares paid by flexible leisure travelers, and (c) unprecedented reduction in travel by both types of passengers as a result of the 11 September incidents.

Making money in the airline business is difficult; always has been, given the nature and complexity of the business, as well as some unusual constraints under which managements work. However, 'business as usual' is no longer a viable option given (a) the increasingly seriousness of events (for example, the impact of the 11 September incidents vs. the impact of the Gulf War), (b) the speed at which the industry is being affected by the convergence of powerful forces—for example, globalization; customer expectations and power; and technology, and (c) the unacceptability of the financial performance for shareholders. Consequently, there is a critical need within the industry for proactive and strategic management to produce customer and shareholder values in today's rapid pace of economic, demographic, social, political, and technological change. While proactive and strategic management is important for all airlines, it is imperative for the second-tier airlines since we cannot count on the continuation of the three-tier structure mentioned above.

Technology is probably the single most important force affecting the airline industry. It is not only a driver of change, but at the same time it can also be a resource for the industry to manage proactively the needed fundamental structural change. Technology is being adopted at an increasingly rapid rate. Consider the number of years it took a certain percentage of households to get a telephone, a television, a computer, and access to the Internet. The rapid speed of adoption of specific technologies and trends is becoming more critical, and it has many implications for smaller and larger airlines.

In the past, various developments in technology provided airlines the opportunity to reduce costs, enhance revenue, improve customer service, or some combination of all three. Improvements in aircraft technology, for example, enabled airlines to reduce operating costs that they decided to pass on to passengers and shippers in the form of lower fares and cargo rates to generate growth. Improvements in information technology and communications played a key role in the development of computer reservation systems that dramatically changed the mode and cost of distribution of the airline product. Emergence of the Internet and wireless communications enabled airlines to develop comprehensive websites, electronic ticketing, online check-in facilities, airline-linked mobile services, and a broad spectrum of in-flight entertainment and services.

Some technologies enabled airlines to reduce significantly their operating costs. Generally, however, airlines did not benefit fully from the reduction in their operating costs. Often, they either enabled other members in the value chain to reduce their costs or airlines decided to reduce prices and generate profitless growth on the back of price elasticity. Airlines also used technology to focus on optimizing their production and operations with respect to costs or some limited resources such as fleet, airport slots, or bilateral authority instead of focusing on the understanding and optimization of customer satisfaction. Such a focus resulted in common product features and similar cost structures for traditional incumbent carriers, providing an opportunity for new entrants to offer a different price-service option. Some technologies could have enabled airlines to simplify their business processes by re-engineering their products and services. The incumbent airlines did not or could not take advantage of the cost savings—for example, unwillingness to change organizational culture, inflexible labor contracts, and constrained infrastructure—providing, once again, an opportunity for new entrants with a different business model.

Some technologies enabled airlines to enhance their products or services—tangible features such as the seat itself and in-flight entertainment systems or intangible applications such as customer relationship management and compressed cycle times. While such innovations did for some airlines create a first mover advantage through increased share of value-generating business, such strategies and actions often increased the cost base for the entire industry while generating profits for the vendors of these technology products and services. Technology did not provide long-term competitive advantage for airlines that developed and used technology. Treating technology as a strategic imperative, most competitors followed the leader, not so much to increase shareholder value, but more to protect it.

One basic conclusion from the preceding observations—and, admittedly, generalizations—is that the airline industry needs to have a better business discipline to capitalize on emerging technology to generate sustainable shareholder value for *itself*. Airlines can create traditional and non-traditional competitive advantages by carefully selecting and investing in technologies. It was pointed out in Chapter 2 that the ability to accurately measure the profitability of an airline by product and by customer can enable an airline to re-focus resources and create a new category of sustainable competitive advantage—as good as if not better than the traditional low cost, product leadership, or operational excellence drivers of competitive advantage. Similarly, it was pointed out in Chapter 7 that proficient management of intangible assets such as customer and employee relations, and timely and actionable information can lead to a non-traditional source of competitive advantage, such as a skilled workforce and a value-based planning capability. Some analysts claim that it is difficult for airlines to innovate on their own and that synergistic alliances are needed within the 'ecosystem'—airlines and other providers of services such as airports and government agencies. Technology is available to support management in areas related to all three examples—accurate measurement of profitability at the granular level, management of intangible assets, and the development of synergistic alliances within the ecosystem.

The problem is not the lack of technology; it is the intelligent selection of, investment in, and implementation of value-adding technology. Figure 8.2 shows an illustration of investment/integration requirements vs. value of selected technologies. There is no question about the need of airlines to select technologies that adapt to the demands of the customer, but they must also adapt to the needs of the business—both employees and shareholders. Take, for example, the investment in in-flight entertainment systems. Initially, airlines charged for this service and considered the investment in technology to be a source of revenue. In reality, one-by-one they began to provide the service free and technology became a cost item for the industry rather than a source of revenue.

Customers are going online. Therefore, airlines have begun to go online to provide reservation and electronic ticketing capability and wireless application protocol technology to provide real-time information on the status of flights. Online technology is providing customers with anytime and anywhere access to the airline system. However, airlines must also focus on the use of online technology to move from reactive marketing to real time proactive marketing to enhance revenue and, for example, to

A Call for Action 189

implement e-learning to improve employee productivity and the ability to provide more valuable service to passengers and shippers. Smaller and developing airlines can use online technology to enable themselves to compete with the larger developed airlines by acquiring and using sophisticated network planning systems over the Web.

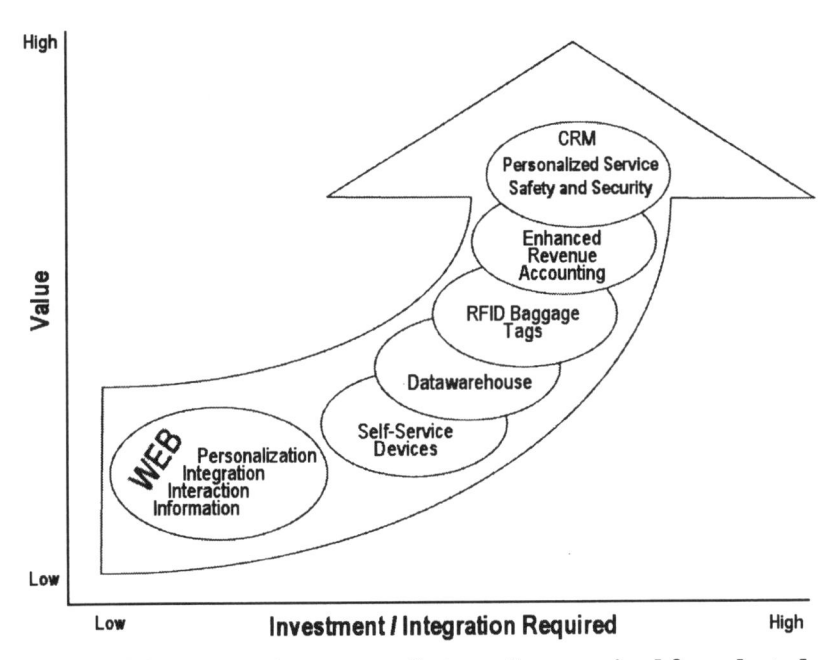

Figure 8.2 Value versus investment/integration required for selected technologies

As shown in Figure 8.2, airlines can begin with the development of cost-effective websites. But, even here, each airline must choose the level of investment—ranging from websites that provide only the basic and generic information to those that provide personalized and interactive information. At the other end of the spectrum, technology is available to enable an airline to provide—and, hopefully, charge for—truly personalized service. However, investment and integration costs can be high. Airlines must therefore analyze carefully the value of a particular technology (from the perspectives of customers and businesses) as well as the total costs of its acquisition and implementation. Implementation of some technologies requires a different mindset to change organizational structure, processes, and culture—changes that can be expensive and risky. Moreover, costs of implementation, integration, and, above all, day-to-day maintenance can be high. Such prerequisites are clearly required to derive the full benefit of CRM and e-business initiatives presented in Chapters 3 and 4. On the other hand, the previous chapter also discussed the value of basic technologies—such as enhanced revenue accounting (fare audit and proration) to improve revenue—that require virtually no change in the organization's structure or culture.

One of the most cost-effective investments in technology shown in Figure 8.2 is the development of a data warehousing and mining capability. The lack of information about customers and products is a key problem with the development and delivery of products that have value both for the customer and the business. And even if the information is available, it exists in fragmented databases: some exists in the computer reservation system, some in the sales division, some in the complaint department, some in the frequent flyer programs, some in the lost baggage department, some in the finance department, some in the operations department, and so forth. Valuable are the customer touch points where information is not being captured. Consider, for instance, the amount of information available from flight attendants on business-class and first-class passengers traveling on long-haul flights.

Airlines typically do not have a unified view of a customer or a product. First, if an airline wants to become customer-centric—meaning that it is capable of knowing and understanding its customers—it must develop a comprehensive data warehouse (and data mining capability) that integrates the information contained in different databases. In any case, such a unified view of customers is needed to implement higher-level technologies such as CRM and value-based strategies.

Second, if an airline wants to become shareholder-centric, comprehensive, valid, and timely information is needed on customer and product profitability. Improved data and analyses enable an airline to understand profit in different segments so that management can invest only in aircraft capacity that is likely to enhance shareholder value. It is the inability and/or the failure to do such analyses that have led many airlines to invest in profitless growth or even value-destroying growth—actions that have been damaging for individual airlines as well as for the industry as a whole. The ability to conduct such analyses at the executive level

A Call for Action 191

requires the availability of sophisticated business intelligence systems that can assist senior executives in providing improved business control at the strategic level from the information obtained from operational systems. Figure 8.3 shows an example of a business intelligence/knowledge management system that can identify frequently asked questions, key performance indicators, and benchmark indicators. This figure is a basic representation of the IATA CEO Cockpit project that can assist airline CEOs and senior functional executives to focus on the airline enterprise-wide information to proactively manage the airline. While many airlines have management information systems that support individual business areas, the CEO Cockpit initiative is a knowledge management and business intelligence system that cuts across the whole airline enterprise and turns data into actionable information to make informed customer-centric and business-centric decisions.

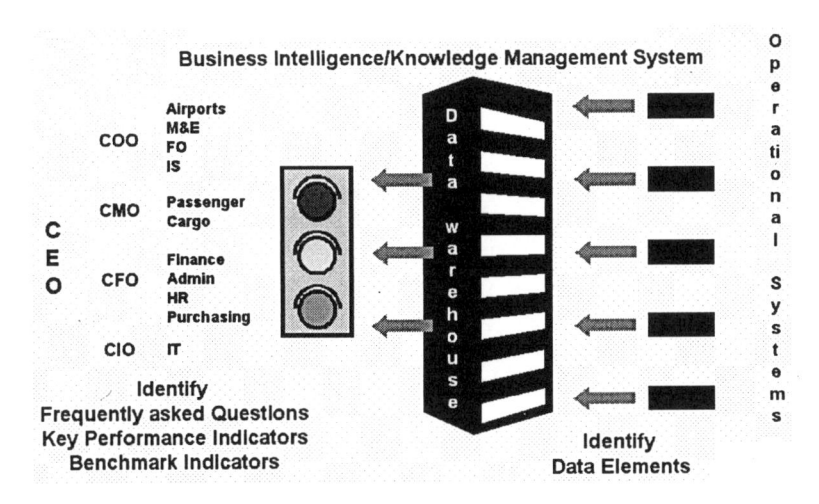

Figure 8.3 CEO cockpit executive information system
Source: International Air Transport Association

Having access to the appropriate information is, however, the easy task. Using the information wisely—such as cross-functional optimization and collaborative approach to planning and execution of strategy—is a more difficult task. An even more difficult task is having the courage to act on what the information dictates to make vital decisions, such as the decision to 'fire' the unprofitable segment of customers, leading to a significant reduction in the size of operations.

Knowing that emerging technology is available for airline managements to drive the long-overdue transformational business strategies is one thing; making intelligent investment in the right technologies coupled with making the prerequisite changes in the mindsets, organizational structures, and business processes to successfully implement these technologies is completely something else. It is important to keep in mind that it is more valuable to use technology to plan for an airline than to save an airline. In this context, hopefully, some observations made in this book will help practitioners in the airline industry to align judiciously their business and technology strategies to proactively design a value-generating business. The reader may not agree with the role that emerging technology will play in driving and enabling airline business strategies, as presented in this book; but, hopefully, some of the points raised will stimulate the reader to think about the use of emerging technology to understand and conquer the new and diverse business environment and, at the same time, create shareholder value.

Index

Airbus
   A380, 40, 109-14, 147
   Corporate Jetliner 46, 47
aircraft
   configuration 38, 41, 48, 113, 114
   families 29
   fractional ownership 123-6
   freighters 145, 146
   personal 122, 126-9
airline
   business travelers 47, 65, 116, 186
   changing dynamics 14-25
   competition 14-17
   costs 50, 51, 52, 53, 54, 95, 115, 116, 126, 154
   customer 1, 10-14, 86
   e-business 94-6
   employees 97, 98, 107, 119, 178, 179, 180
   fares 6, 11, 14, 15, 39, 40 52, 68
   fractional ownership 123-6
   leisure travelers 67, 68, 186
   low-fare 15, 16, 17
   partners 96-97
   product development 37-50 64
   profitability 30, 31, 50-54, 60, 185, 188
   scheduling 29, 30
airport processing 42-6

biometrics 19-20, 43, 49, 156, 162
Boeing
   Business Jet 47, 124
   Sonic Cruiser 115-7
Bombardier 47, 118
broadband 19

cargo
   aircraft 145-7
   belly 133, 136, 140, 146
   charter airlines 141, 142
   consolidation 137
   forwarders 137, 140, 141, 142, 144
   freighters 138, 145, 146
   growth 131
   industry structure 136-45
   information management 147-150
   integrators 140, 141, 142, 143
   postal services 141, 142, 143
   products 132
   service criteria 134

shipper needs 131-6
strategic alliances 137, 138, 139, 140
time-definite 133, 144, 147
yields 132
CIO 168-70
Cirrus SR-20 128, 129
crew scheduling 30, 32, 33
CRM 32, 71, 72-84, 149, 157, 161, 162, 189
cross-functionality 28, 30, 32, 34, 36, 37, 165
customers, *see* airline, customer
customer expectations, *see* airline, customer
customer satisfaction 98; *see also* airline, customer

data
bases 44, 51, 75, 78, 102,171-5, 190
mining 75, 82, 190
warehousing 82, 190
delays 12, 30, 33, 34, 35, 104, 166, 173
demography
Asia-Pacific 6, 7
baby boomers 58, 59
Central Europe 7
cultural aspects 21-3
generation X 58
Latin America 6, segmentation 58, 59
Western Europe 7
world populations 1-9
developing countries
air travel 5
populations 2

distribution channel 61, 62, 91, 105, 154

e-business
constituents 92
description 88
effects 92
key drivers 99-108
metrics 105-6
processes 104-5
technology 100-104
transformation process 106-8
value of 89, 90
vision 100
Eclipse 500 126, 127
e-commerce 87, 95, 149, 150, 172
e-learning 180
experience economy 18

Fairchild Dornier 118, 120, 121
frequent flyers 20, 49, 59, 60, 61, 163

hubs 29, 51, 112

IATA 43, 44, 48, 49, 114, 162, 191
in-flight entertainment 41, 49, 103, 187, 188
information
economy 18
reach 19
richness 19
intelligent services 12
Internet 13, 17, 18, 19, 22, 42, 43, 95, 147, 148, 150, 158, 172, 181, 186

irregular operations 34

knowledge management 176-82

legacy systems 103, 148, 170-72
loyalty programs 13, 14, 61

m-commerce 100
market
   fragmentation 112
   segmentation 57-72
mobile communications 13, 19, 22, 23, 100, 101, 155, 158

NetJets 124, 126

operational planning 28-37
outsourcing 139, 167

passenger satisfaction 98; *see also* airline, customer
personal transportation 122-9
populations, *see* demography
profitability, *see* airline, profitability
product development 37-50, 64

real time
   operations 29, 30
   connectivity 100
regional
   jets 118, 120
   markets 118-21
   travel 5

RF chips 44, 160
revenue accounting 159
revenue management 29, 31, 51

safety and security 20-21
scheduling 28, 29, 30, 32, 37;
scope clauses 120
segmentation, 31, 57-72, 185
segment evaluation 68, 69
self-service 42, 92, 156
simplifying passenger travel 43-46, 49
smart card 22
Sonic Cruiser; *see* Boeing

technology
   role of 154-163
travel
   card 20, 42-5
   device 44
Travel Research Center 63 64, 66, 67, 70

u-commerce 100

value-based planning 163-68
video conferencing 5
virtuGate 35, 36
voice recognition 19

Web 16, 22, 96, 100, 102, 103, 106, 149, 154
websites 16, 23, 43, 94, 103, 135, 155
wireless communications 22

About the Author

Nawal Taneja has more than 30 years of experience in the airline industry. As a practitioner, he has worked for and advised major airlines and airline-related businesses worldwide in the areas of strategic and tactical planning. His experience also includes the presidency of a small airline that provided schedule and charter service with jet aircraft and the presidency of a research organization that provided consulting services to the air transportation community worldwide. In academia, he has served as Professor and Chairman of the Aerospace Engineering and Aviation Department at the Ohio State University, and a faculty member of the Flight Transportation Section of the Massachusetts Institute of Technology. On the government side, he has advised civil aviation authorities in public policy areas such as airline deregulation, air transportation bilateral agreements, and management and operations of government-owned airlines.